EVERGREEN ISLANDS

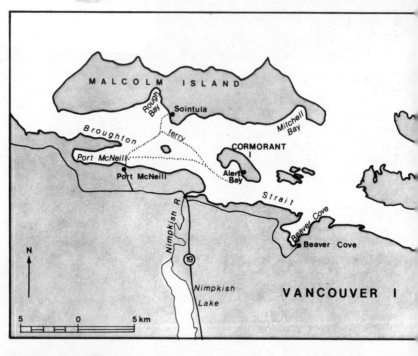

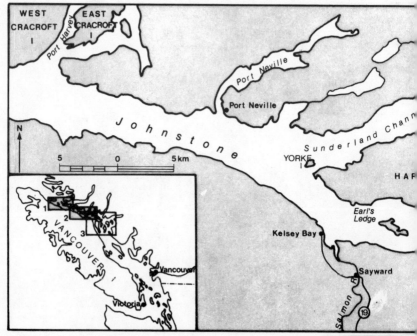

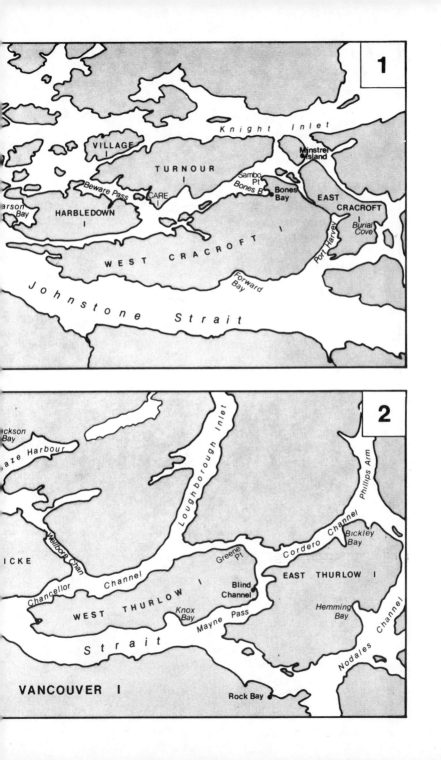

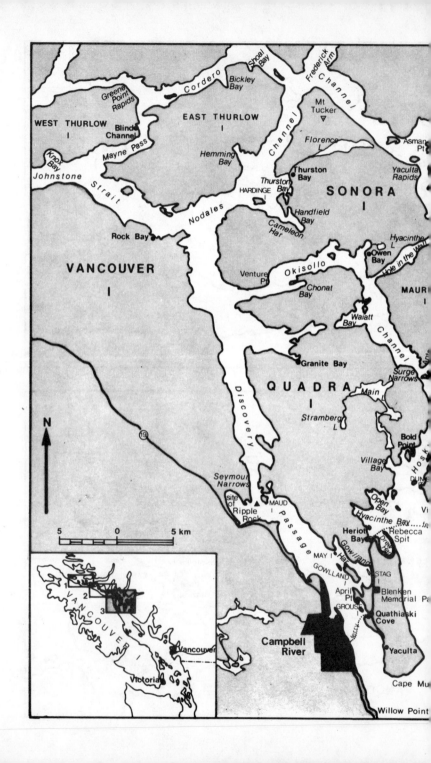

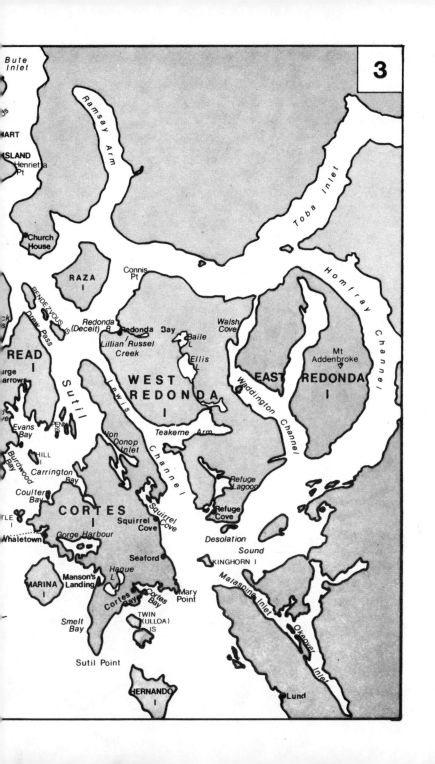

3

Bute
Inlet

Ramsay Arm

Toba Inlet

JS

MART

ISLAND
Henrietta
Pt

Church
House

Homfray Channel

RAZA
I

Connis
Pt

BENDEZVOUS
IS

Draw Pass

Walsh
Cove

Redonda
(Deceit) R

Redonda 3ay

Baile
L

READ
I

urge
arrows

Lillian
Creek

Russel

Mt
Addenbroke

WEST
REDONDA
I

EAST   REDONDA
I

Ellis
L

Sutil

Lewis

Waddington Channel

Evans
Bay

PEN
IS

Teakerne Arm

HILL
I

Von
Donop
Inlet

Channel

Burdwood
Bay

Carrington
Bay

Coulter
Bay

Refuge
Lagoon

TLE
I

CORTES
I

Squirrel
Cove

Squirrel
Cove

Refuge
Cove

Whaletown

Gorge Harbour

Desolation
Sound

Seaford

KINGHORN I

MARINA
I

Manson's
Landing

Haque
L

Malaspina Inlet

Cortes
Bay

Cortes
Bay

Mary
Point

Smelt
Bay

TWIN
(ULLOA)
IS

Okeover Inlet

Sutil Point

HERNANDO
I

Lund

EVERGREEN ISLANDS

THE ISLANDS OF THE INSIDE PASSAGE:
QUADRA TO MALCOLM

Doris Andersen

GRAY'S PUBLISHING LIMITED, SIDNEY, BRITISH COLUMBIA

Typesetting by The Typeworks, Mayne Island,
British Columbia

Cover photograph courtesy of British Columbia Forestry
    Service, Victoria
Cover design by Nicholas Newbeck Design, Victoria
Maps designed and drawn by Peggy Ward, North Vancouver

Canadian Cataloguing in Publication Data

Andersen, Doris, 1909–
        The evergreen islands

Includes index.
ISBN O-88826-084-9
1. Inside Passage-History. I. Title.
FC3845.I57A53   971.13   C79091065-9
F1089.I5A53

CONTENTS

ACKNOWLEDGEMENTS

The Campbell River and District Museum supplied valuable information for my research, making available its photographs, tapes and clippings, largely collected by John Ackroyd of the work projects Genesis I and II. Curator Jay Stewart and assistant curator Jeanette Taylor and their predecessors, John Frishholz and Norma Holmes, were especially helpful. Historical photographs, manuscripts, letters and news articles were also obtained from the Provincial Archives at Victoria, with special help from Barbara McLennan of the Visual Records Division. Help was also received from the Provincial Museum's Ethnology Department, the Provincial Library and Legislative Library, the Public Archives of Canada in Ottawa, J.E. Forester of the British Columbia Department of Mines, D.J. Brant of the Department of Indian and Northern Affairs in Ottawa, the Office of the Regional District of Comox-Strathcona and the Vancouver Public Library's Northwest Department.

My thanks go to Harold Griffin, editor of *The Fisherman*, for access to back files and its picture file, and to Quadra School for access to its archives file. I am grateful to editor Sophie Gregg for back copies of *Discovery Passage*, to Clara Vaughn for showing me her many scrapbooks of newspaper clippings about Quadra Island, and to J. Proudfoot of the Correctional Branch for information on the Redonda Bay camp. Anthropologist Joy Inglis acted as my very knowledgeable guide to the site and midden at Tsqulotn. My thanks go also to Marion Antle Mennes of New York, Alan Greene Jr. and the daughters of R.J. Walker for permission to use material from the writings of their parents.

Newspaper articles published over the years are too numerous to credit by writer, and were located through copies in various archives or through indexes of newspaper microfilms in libraries and archives. Several were mailed to me by islanders previously unknown to me, whose kindness is deeply appreciated.

The many residents of the islands who supplied information include Harry Assu, Arthur Bendickson, Edith and Harold Bendickson, Albert Bigold, Bert and Mary Brimacombe, Arthur Calwell, Della Chellew, Arthur Clandening, Helen and Gus Clements, Mrs. Dan Cranmer, Doug Cranmer, Doris Davies, Bob Davis Sr., Marre Dodman, Gilean Douglas, Bob Early, Isabelle Edgett, May Ellingsen, Monica Fremont, Chester Haas, Tom Hall, Louise Hanes, Walter and Mary Hayes, Louise Hovell, Alice and Eddie Joyce, Ture Krooks, O.F. Landauer, Bill Law, Mary and Oscar Lewis, Rose McKay, Alex Mercier, Frank and Jean Mowat, Bruce and Kay Noble, Eiva and Harold Noble, Martin and Alan Olmstead, Dollie Pidcock, Alvina Poitras, Ethel and Harold Redford, Ella and Jack Schibler, August Schnarr, Gwen Sparks, Joseph H. Wales, Bill Whittington, Audrey Wilson, and Reg and Marie Wilson.

INTRODUCTION

The Islands

The story of the islands of the Inside Passage that lie to the north of the Gulf of Georgia and extend into Johnstone and Broughton Straits involves a kaleidoscope of colourful tales of sturdy pioneers who braved wild rapids at sea and wild animals on land, of outlaws and violent deaths, of piratical raids, and of men and women whose stubborn determination produced fertile farmlands hacked from virgin forest.

Quadra Island, largest of the group and dealt with here in greatest detail, has an eventful recorded Indian history, from Captain Vancouver's description of the Salish village, through the Kwakiutl invasion, down to its present modern village. Cortes, with its dramatic and varied scenery, is still largely unspoiled by development and retains an aura of romance from its early whaling history. Read Island, with a steep, rocky shoreline, surprises with a few farms on the low-lying land, but was essentially an island of independent individuals, of loggers, fishermen and lovers of a wilderness life, an island with a flamboyant past of outlaws and violence. Malcolm Island, last on the list, is the story of Sointula, a search for Utopia, a tale of inspiration and of heartbreak.

All throughout the waters surrounding the larger islands lies a myriad of other islands of all shapes and sizes; some mountainous, some low-lying, some rocky, others heavily timbered. They moved Marcus Smith in 1872 to write of the "extraordinary intricacy and surpassing beauty of the Archipelago... islands of infinite variety of size and form... bold headlands and cozy bays, deep and narrow channels leading into romantic and snug harbours..."[1] These northern islands are rapidly becoming known to and sought by tourists and settlers. For them, and for those who have long known and loved the islands, this history is written.

CHAPTER 1

Quadra Island

"... tides you've never seen
Assault the sands of What-has-been,
And from your island's tallest tree,
You watch advance What-is-to-be. ..."
                                   -Edna St. Vincent Millay[1]

The first glimpse of Quadra Island, for those travelling northward, shows the steep cliffs of Cape Mudge; then comes the slim white shaft of the lighthouse with its adjacent buildings, and farther along a little cluster of houses skirts the waterfront, showing up as pink, yellow and white smudges against a background of thick evergreen forest. Beyond these, a fleet of trim seine boats lies at anchor. Nothing about the peaceful scene suggests that only a hundred years ago the area was known as the death-hole, haunt of piratical raiders who struck terror into the hearts of those white men and Indians forced to pass the Kwakiutl village as they plied the narrow waterway of Discovery Passage.

Quadra, largest of the Discovery Group, is nestled between massive Vancouver Island and the smaller islands of Sonora, Maurelle, Read and Cortes. Cape Mudge is at Quadra's southernmost tip, and it was here, in 1792, that Captain Vancouver, exploring the coast in his search for the Northwest Passage, climbed the steep cliffs to the Indian village of Tsqulotn and wrote his description of its then-Salish inhabitants.

After an invasion by the Lekwiltok tribe of Kwakiutl Indians drove out the Salish, Quadra experienced widespread changes in lifestyle. The most dramatic is the story of the aggressive Lekwiltok at Cape Mudge, who abandoned their fierce forays and centuries-old culture and embraced an entirely new way of life. The transformation took place during the life of Chief Billy Assu, and at the time of his death in 1965, the Wiwekae band of the Lekwiltok (their name is derived from the plural form of *Wekae*, their mythical ancestor) had proven their ability to adapt and yet remain a close-knit, separate community.

It was an invasion of loggers and miners in the latter part of the nineteenth century that altered the island from an unsought wilderness to a lively centre of industrial activity. Why did they choose Quadra over other northern islands? For one reason, timber stands were mature and numerous bays were available for safe booming. Also, the assay reports of the first miners proved the presence of valuable minerals in many areas of the island. The industries brought in workers; the workers brought in families. As population increased, Quadra acquired hotels, post offices, public schools, a mission, and the only policeman and jail serving the surrounding area, while Campbell River across the passage was still unsettled forest land.

Settlers and industries on Quadra concentrated in Heriot Bay, Quathiaski Cove, Granite Bay and Gowlland Harbour. When the industries petered out some years later and the sawmills burned down, the forest reclaimed the abandoned mines and logging camps and the population decreased to the few who valued an independent, rural life. Settlers who joined them over the years were those with similar tastes who sought escape from the pressures of city living.

During the depression years of the 1930's, newcomers were mainly the unemployed, who came to squat along Quadra's southern coastline where the abundant salmon and shellfish kept living costs low. As Campbell River's population grew and its industries multiplied, Quadra's economic dependence upon this town increased, and many residents moved to jobs across the passage.

Population figures rose again on Quadra with the acquisition of Rebecca Spit by the government, its opening to the public as a provincial park in 1959, and the purchase of the first car ferry between Vancouver Island and Quadra the following year. In the resultant tourist invasion, population figures more than doubled during the summer months, and the island was faced with many of the problems already encountered by the southern Gulf Islands.

Quadra is the second largest and one of the more important of all the inside islands. Twenty-two miles long and one to ten miles in width, it has a resident population judged in the 1971 regional planning study to be around 1300, topped only by Saltspring and Texada.

Visitors to sheltered Drew Harbour or to Heriot Bay's island-encircled waters, smooth as a placid lake in calm weather, might well remain unaware of the dangers surrounding the island,—dangers that beset early voyagers and that lie in wait for travellers today who ignore tidal charts and storm warnings. Rapids and rough waters form a barrier to the island when southeasterly storms and flood tides coincide. Over the years, Seymour Narrows to the west has been a constant threat to navigators. (" . . . woe be unto the small boat which gets into its mighty grip," warned the *British Columbia Directory* of 1892.) A description of the rapids, which were caused in part by massive submerged rocks, was given by Cap-

tain Vancouver in 1792. "The tide, setting to the southward through the confined passage, rushed with such immense impetuosity as to produce the appearance of falls
considerably high; though not the least obstruction of
either rocks or sands, so far as we had an opportunity of
examining it, appeared to exist."[2]

Originally, the narrows were called "Yaculta" after the
equally fierce Euclataw or Lekwiltok Indians. When they
were renamed in 1846 for Sir George Seymour, the term
"Yaculta" was transferred to the rapids near Stuart
Island.

The northeastern shores of Quadra are fringed by a
row of large and small islands that form narrow channels,
such as Upper Rapid and Surge Narrows, agitated by
turbulence and whirlpools during flood and ebb tides. At
the foot of Cape Mudge's 200-foot high cliff a stony reef
extends, creating a dangerous tide rip. Lieutenant
Mayne, R.N., described Cape Mudge in 1860: "This part
of the Gulf of Georgia forms a sort of playground for the
waters, in which they frolic, utterly regardless of all tidal
rules."[3]

Cape Mudge has a long history of shipwrecks off its
shores, despite the erection of a lighthouse in 1898.
Scotsman John Davidson acted as lighthouse keeper for
20 years, using a hand-pumped fog horn to give a feeble
warning to passing ships when mists or snowstorms cut
visibility to the danger point. Ships have foundered and
sunk within minutes when venturing into these wild
waters at times when stiff southeasters blow and the
surging tides, flowing from north and south, collide at
the cape. The 65-foot tug *Petrel* vanished there in the
1940's, leaving no trace behind.

The removal of Ripple Rock in 1958 did much to tame
the ferocity of Seymour Narrows, and Cape Mudge is
harmless in calm weather during slack tide. The beauty
of Quadra's sheltered bays, with their outlook on green

forested islands backed by snow-capped mountain ranges, remains a magnet to draw settler and tourist despite the menace of the island's encircling rapids.

-2-

Two names given to the Island caused confusion for well over a century. During this period the area was never adequately charted, and three adjacent islands were thought to be one large land mass, drawn as such on Vancouver's and Galiano's charts and called Valdes Island in honour of the Spanish explorer Cayetano Valdes. Valdes and Galiano had met the English explorer in the Strait of Juan de Fuca, and the two parties joined forces to explore the northwest coast and its islands.

Captain Richards of H.M.S. *Plumper*, who charted the southern Gulf Islands in 1859, gave one of these the same name, Valdes, after which the northern Valdes was usually referred to as Big Valdes. Sandford Fleming, Engineer-in-Chief of the survey for a possible Canadian Pacific Railway route to the Pacific in 1872, refers to "the group of Islands known as Valdes Islands",[4] and Gosnell's 1897 yearbook refers to "Valdez" Island as lying at the entrance to Bute Inlet. (The island referred to is today's Maurelle Island.)

Eventually, Big Valdes was properly charted and in 1903 the three islands composing it were renamed Quadra, Sonora and Maurelle by the Canadian Board on Geographic Names. However, both the southern Gulf Island near Nanaimo and northern Quadra were still labelled "Valdes" and "Valdez" on the Department of the Interior's 1909 map, and for the next 30 years many travellers and writers continued to call Quadra by its early name.

Quadra was chosen as Big Valdes' name because Vancouver Island had discarded Quadra from its original title,—the Island of Quadra and Vancouver,—and it was

thought fitting for the name of Vancouver's good friend
to reappear on an island lying close beside it, across Discovery Passage.

The historic meeting of Captain George Vancouver
and Don Juan Francisco de la Bodega y Quadra
recorded by the explorers almost two centuries ago
retains its charm and humour to this day. An instant
friendship sprang up between these two men, dissimilar
in character and unable to speak each other's language.
When the time came for them to part, Vancouver wrote
in his journal that it was a "painful consideration" to
think they might not meet again.

Quadra had come to Nootka in the spring of 1792.
Vancouver arrived at Friendly Cove in August of that
year and went to pay his respects to Quadra, taking with
him Thomas Dodson, mate on the British ship *Daedalus*,
who spoke Spanish and acted as interpreter. Vancouver
was a just but stern disciplinarian who frequently sentenced his crew to the lash and dealt severely with
Indians who stole articles from his ship. Quadra was
easy-going and generous with Indians and whites;
Vancouver deplored the Spaniard's "mildness and lenity"
in matters of discipline. "Mr. Quadra was too good a
man," said Mr. Bell of the *Chatham*, consort of the *Discovery*. "He even treated the Indians more like
companions than people that should be taught
subjection."[5]

On the return trip from a joint excursion to visit Ma-
quinna, Quadra "very earnestly requested" Vancouver to
name some port or island after both of them "to commemorate our meeting and the very friendly intercourse that
had taken place and subsisted between us." Vancouver
recorded: "Conceiving no spot so proper for this
denomination as the place where we had first met, which
was nearly in the centre of a tract of land that had first
been circumnavigated by us . . . I named the country the

island of Quadra and Vancouver; with which compliment he seemed highly pleased."[6]

"Quadra and Vancouver" proved too large a mouthful for travellers and too long a title for mapmakers. One of the early charts to omit the name Quadra was the British Admiralty chart published February 28, 1849, entitled "Vancouver Island and Gulph of Georgia, from the Surveys of Captain G. Vancouver, R.N., 1793, Captains D. Galiano and C. Valdes 1792, Captain H. Kellett, R.N. 1847." When the northern Valdes Island was found to be not one but three islands, the name of Quadra was returned to the coast in 1903.

English and Spanish explored the mainland coast in June and July 1792. When they reached Desolation Sound, Vancouver was unsure if their present route would lead to the sea or if they would be forced to backtrack to Juan de Fuca Strait and sail up Vancouver Island's west coast to reach Nootka. To find out, he sent Johnstone ahead in the *Chatham*'s cutter, along with Swaine in the launch, to continue north up the passages ahead, and his officers Puget and Whidbey to explore the east coast of Vancouver Island.

Johnstone battled the Yaculta Rapids to Johnstone Strait and eventually found the sea, while Puget and Whidbey went south as far as Savary Island where they camped and enjoyed the "fine smooth beach" before crossing to Vancouver Island. They sailed west and north until they reached Discovery Passage, anchoring across from Cape Mudge on July 3, 1792. A number of natives from the Indian village on the bluff quickly surrounded the boats with their canoes, and the nervous sailors fired several shots from their swivel gun to discourage them. The Indians paddled away, showing no resentment, and gave them no trouble during the night.

The strong tides in the passage convinced Puget and Whidbey that it must lead to the sea, and they hastened back to report this to Vancouver. When Johnstone's party returned, the combined information suggested the safest and speediest route would be by way of Discovery Passage. Vancouver headed at once for the passage, anchoring off the cape, which he named after his first lieutenant, Zachary Mudge.

Captain George Henry Richards, surveying in H.M. paddle sloop *Hecate* around 1862, named Heriot Bay on Quadra after Frederick Lewis Maitland Heriot, descendant of the sixth Earl of Lauderdale and kinsman of Rear Admiral Sir Thomas Maitland. Around the same time he named Gowlland Harbour and Gowlland Island on the opposite side of Quadra after John Thomas Gowlland, R.N., second master of the *Hecate*.

Captain Daniel Pender also served on the *Hecate*, and after Captain Richards sailed her back to England in 1863, Pender continued the surveys in the paddle-wheel steamer *Beaver*, hired from the Hudson's Bay Company. Around 1867 he named Drew Harbour on Quadra after Charles Randolph Drew, R.N., assistant secretary to Rear Admiral the Honourable George F. Hastings, C.B., flagship *Zealous.*

-3-

Cape Mudge is one of the few spots visited by Vancouver which is easily identified, and where the explorer landed and wrote a detailed description. He says in his journal that he and his officers went ashore and landed at the foot of the steep, sandy cliff, described as about a hundred feet high and almost perpendicular. The Indian chief met them there and Vancouver gave him several presents, after which he led them along a narrow path winding diagonally up the cliff to the village. The houses were built close to the edge of the precipice, strategically placed for defence, and in front of these sat the villagers, who urged the explorers to refresh themselves with fish, berries and water.

Archibald Menzies, the ship's botanist, described the village as consisting of about 12 houses or huts planked over with large boards, some ornamented with paintings on the house fronts. The dwellings were flat roofed and quadrangular and each contained several families. He estimated the inhabitants to number about 350, judging from the 88 large canoes in the water and on the beach.

Menzies gives a colourful picture of the villagers: "Like the generality of Natives we met in the Country these were of a middling stature and rather slender bodied, of a light copper colour; they were awkward in their motions & ill formed in their limbs which no doubt in some measure proceeded from their constant practice of squatting down on their heels in their posture of setting either on Shore or in their canoes. They have flat broad faces with small staring eyes; their Teeth are small and dirty; their Ears are perforated for appending Ornaments either of Copper or pearly Shells; the Septum of the Nose they also pierce & sometimes wear a quill or piece of tooth-shell in it; their Hair is streight black & long, but mixed with such quantity of red-ocre grease & dirt puffed over at times with white down that its real colour is not easily distinguishable; they have long black Beards with long Hair about their privates, but none on their Breasts or on the Arm pits. Some had ornamented their faces by painting it with red-ocre, sprinkled over with black Glimmer that helped not a little to heighten their ferocious appearance."[7]

Menzies wrote of a walk through the dense forest behind the village and westward along the side of the channel, where they came to a level pasture stretching for nearly two miles. They decided this pleasant area must be used by the Indians for games and amusement. While walking there, they saw the native graveyard over to one side of the pasture close to the woods. The dead had been put into small square boxes, or wrapped up in mats or old garments into square bundles and placed above ground in closely boarded tombs.

Some writers have called this the first recorded description of Kwakiutl Indians, but it is known now that the cliff-top village of Tsqulotn was Salish, *Tsqulotn* being the Salishan word for "playing field." Menzies said the amiable natives spoke a broken dialect much like that of the Salish natives of the east side of the Gulf of Georgia. Also, the Kwakiutl put their dead in the branches of trees, an oddity that Menzies would have noted if the burial ground had been Kwakiutl.

It was after Vancouver's visit, around 1850, that a branch of the Southern Kwakiutl, the Lekwiltok, swept down from Jackson Bay in their great war canoes and captured Salishan villages on Quadra in the northern area now known as Granite Bay and in the west on Gowlland Harbour, burning the villages and slaughtering the inhabitants. The cliff-top village of Tsqulotn on Cape Mudge held off the attack, but after the Lekwiltok settled on flat shoreland just north of Tsqulotn, on the site of the present village of Yaculta, the Salish grew uneasy with these fierce warriors on their doorstep. They abandoned their village and moved down the coast to join the Salish in Comox. Today, one can stroll over the same level pasture where Vancouver's party walked, high above the waters of the cape, and see the great midden behind the site of the ancient Salish village. It is all Lekwiltok property now.

The capture of Quadra Island gave the Lekwiltok a splendid vantage point for forays on canoes plying their way through the narrow channel. Somewhat later, more Lekwiltok, the We-wi-kum, this time from Greene Point Rapids and Phillips Arm, settled across the passage on the Campbell River spit. From their two villages, the Lekwiltok were now in complete command of the waterway and shot forth, armed with Spanish muskets, to demand toll or loot from all who travelled past. Great or small, no craft was safe from them. E.S. Curtis quotes one old Indian as saying the Lekwiltok were "like a great mouth, always open to swallow whatever attempted to pass... if one had a diver's suit and were to search the little bays between Cape Mudge and Salmon River, the charred remains of a surprising number of schooners would be found."[8]

Lieutenant Mayne, R.N., described the raiders as "the Ishmaelites of the coast, their hand being literally against every one's and every one's against them."[9] He said several battles had taken place at different times at or near Cape Mudge, and on one occasion the Lekwiltok murdered nearly all the crew of a Hudson's Bay vessel which had stopped there for water.

Indians from Yaculta village at Cape Mudge attacked and robbed some Chinese in 1860. The British gunboat, H.M. *Forward*, was sent to demand a return of the property but was met with shots and defiant shouts from the stockade. The gunboat opened fire on canoes on the beach and finally upon the stockade, but it was not until several men were killed that the Lekwiltok agreed to hand over their loot. Mayne said the Cape Mudge Indians were not the only troublesome tribe on the coast but they were the only Indians he had seen who continued to fight after the appearance of a man-of-war before their village.

Lekwiltok warriors ventured far afield, scouring the waters of Georgia Strait and Puget Sound for slaves and spoils. "Yacultas are coming!" was a signal for Salish tribes to flee for the woods. One Salish group built their huts close to Fort Langley's stockade for protection, but twice the reckless raiders paddled up the Fraser River to attack them. It is said that in the last assault on the fort in 1837, the river was black with great war canoes each holding 20 fierce, yelling warriors. The Salish fled for cover, but this time the fort opened fire with its cannon, blasting the canoes from the water and killing vast numbers of the invaders. Few escaped as the Salish rushed from hiding to butcher the injured survivors.[10]

The major wars of the Lekwiltok took place in the days of Chief Billy Assu's father, and the ruins of their old wooden stockade could still be seen when Assu was a child. It was a time, he said, of "much trouble... much fighting, much bloodshed." Chief Billy remembered the last war of the Lekwiltok, when invaders from Fort Rupert attacked and were driven off with the loss of eight enemy canoes.

Piracy brought wealth to the Lekwiltok, but in 1862 their raiding rewarded them with a grisly booty. An epidemic of smallpox brought in by a visiting American from San Francisco, had swept through Victoria, killing off one-third of the Indians who were camped on the outskirts of the city to trade for supplies. The surviving Indians were ordered to return home, and the Lekwiltok, darting out into the channel as usual to capture loot, brought back the deadly disease to their village as well.

Aside from their success as raiders, the Lekwiltok were an enterprising, proud, determined people with a talent for seizing every opportunity for profit. Brave to the point of foolhardiness, in defence as well as in attack, they were brought close to annihilation through the introduction of drink and disease. In some cases, whole bands and villages of coast Indians were wiped out. But Lekwiltok is said to mean "unkillable thing", and is also applied to a large sea worm which cannot be killed by cutting it into pieces because the separate parts wriggle away through the water.[11] The Lekwiltok Indians survived the coming of the white man, but only by destroying their own original culture.

-4-

The first white men to approach the island were explorers and traders. They introduced the Indians to many articles, previously unknown to them, which were then eagerly desired for potlatches. Later, the logging companies provided work and wages and the Indians were able to amass enormous quantities of these goods for their ceremonies. Work and wages superseded the more hazardous method of obtaining wealth through raiding parties for slaves and blankets, and slavery gradually died a natural death. Then the missionaries arrived, denouncing the potlatches, and in 1884 the federal government passed the Potlatch Law which forbade them. Throughout the years, while Quadra Island grew and developed, this clash of cultures continued to arouse vexation and pain in white man and Indian.

First to purchase land on Quadra was William R. Clarke, who bought 144 acres on Gowlland Harbour in May of 1882, followed by pioneer settler R.H. Pidcock who purchased land at Quathiaski Cove. Both transferred their acres to William Sayward, owner of a Victoria sawmill. Sayward purchased large areas of land on Quadra as did independent logger and timber buyer Moses Ireland. Then the rush was on, with many large lumber companies, such as Moodyville Sawmill Company, Royal City Planing Mills Company and Merrill Lumber Company of Seattle, purchasing land and timber leases in the 1880's from one end of the island to the other.

The *British Colonist* reported on December 27, 1883 that resentful Indians were hampering intruders on the island by falsifying blazes and destroying witness trees. However, when the logging camps were established, the Lekwiltok proceeded in their usual practical manner to profit from the newcomers. Indian Agent Pidcock announced that he had given permission to the Cape Mudge Indians to cut timber on their reserve for saw-logs, and in 1889 they were making dogfish oil, which they sold at a good profit to the logging camps as grease for the skids.

Problems arose as whisky pedlars found paying outlets in the logging camps and the reserve. Pidcock, writing about a Cape Mudge potlatch in his Agent's Report of 1890, said: "They are the most reckless, but at the same time the most energetic Indians in the agency, and, if they could have more constant supervision and liquor could be kept from them, they would soon become greatly improved, as they have a splendid reserve and a few of them have built decent little houses and made an attempt at cultivation. The logging camps in the neighbourhood, while offering employment, are a great snare."

Quadra Island was called a loggers' paradise in the days of the first settlers. Timber grew down to the shoreline and there were numerous sheltered bays where logs could be skidded to the smooth waters to be boomed and towed away. There was little or no underbrush to battle between the huge trees that were five and six feet in diameter. Loggers flocked to the island in the 1880's; by 1897 the *Yearbook of British Columbia* spoke of "Valdez" as one of the principal timber locations.

Oxen were used to drag out the great logs. King and Casey were logging with oxen near Quathiaski Cove in 1887; there were "bull camps", using oxen, also at Open Bay and Heriot Bay; Hiram McCormick's logging company worked at Hyacinthe Bay and the Hastings Sawmill Company was at Granite Bay. In the summer of 1894 a tugboat was built for logging on one of the lakes in the interior of the island. Later, the Moffat brothers constructed a large wooden dam at the outlet of the lakes to raise the water level so logs could be towed to a long flume and carried to Village Bay. Jack Moffat was fatally injured on the job. He was rushed to Powell River in Seymour Bagot's powerboat but died before a doctor was reached.

The loggers were responsible in many ways for the rapid development of the island. Their roads gave the settlers access to interior areas and encouraged them to clear farmland in a number of different locations. It was primarily to serve the logging camps that the Columbia Coast Mission and the Union Steamship Company started their up-island services, but they also aided settlers and encouraged more to come.

The *British Columbia Directory* gave a glowing picture of Quadra in 1892, saying deer and grouse were plentiful and bears and panthers were no longer on the island. The waters were said to abound in fish and the woods in game, while cattle and hogs roamed at large on the fine pasture lands and became fat and good for market. "Taking everything into consideration," the directory concluded, "this is the best spot on the globe for the poor man to build his house."[12]

Some came because work was available in the logging or mining camps; some stayed on after the camps closed down and became permanent settlers. Others came searching for a place untouched by the noise and feverish competition of city life. Here, among the giant firs, they built their log cabins, cleared their land and farmed, fished in the lakes and sea, and hunted the abundant game in a poor man's paradise.

Their coming resulted in benefits for Quadra that were not found on the surrounding small islands or in the Campbell River area for years to come. Before 1900, the island boasted two post offices, a public school, a Methodist mission, a hotel, lumber camps, mills, and twice-weekly steamer connections. The authorities of the district, now located on Vancouver Island, were all on Quadra in the early days. The constable's home at Quathiaski Cove was also the jail and court for the district until Campbell River took over in 1924. The house may still be seen on the corner of Heriot Bay and Greene Roads, with bars on the small windows of the rooms that served as the jail.

The Pidcock family has owned land on Quadra for almost a hundred years. The charm of the island captured Reginald Heber Pidcock, Indian Agent for the district, during his tours of duty, and he often camped on its shores, sometimes bringing his young sons with him from his headquarters at Alert Bay. In November of 1882 he pre-empted 167 acres of land at Quathiaski Cove, though some ten years elapsed before he moved his wife and seven children to Quadra, crowding them into a small cabin which later became the kitchen of his large home. There are still Pidcocks living on the island.

Reginald Pidcock was descended from a long line of Anglican parsons. A venturesome streak in his nature had brought him from England with unrealized plans to join the gold rush. A handsome, literate man, he was deeply religious, never travelling without his prayerbook, even when battling wind and tide in dugout canoe or rowboat on his journeys down the coast to visit isolated reserves in the far-flung Alert Bay agency. If there was a church within his reach, he attended every Sunday, and if none were available he conducted services himself.

Before he became Indian Agent in 1886, Pidcock homesteaded in the wilderness that later became the town of Courtenay. He built a sawmill and cut logs on his land for a log house and to supply lumber for a church. He sold some of his land at one dollar an acre for the church site, and he hand-carved the pulpit and reading desk. There were also times when Pidcock and his large family supplied the church with its entire congregation.

Problems plagued him during his term as Indian Agent, as he was instructed to enforce the Potlatch Law but denied for some years a jail or constable to deal with those who broke the law. It must have been a relief for him to escape to his land on Quadra. He camped there at least once a month during 1888, walking over from Quathiaski with his three boys to the Indian village to hear the grievances of the Indians, and conversing fluently with them in the trading jargon of Chinook. The Indians wanted all the land from a point just above their post right across the island, which Pidcock said they were not likely to get.

On July 3 he wrote: "Arrived at Quathiaski, my usual camping place at about 10:30... spent the afternoon in the Indian Village. Found a good deal of gambling going on. Just as I was leaving, a sloop came along... not a whiskey trader as I feared." He went over to the sandspit with Sutton (the owner of the sloop), Sutton being "engaged by the society hunting up Indian curiosities."[13]

Once settled permanently at Quathiaski Cove with his wife and children, Pidcock built himself a comfortable home in the English tradition with a flower conservatory, a tennis court and an orchard. He bought more land until the Pidcock family owned the whole of the cove waterfront property by the close of the century. He built a second sawmill, and his sons logged their land and worked in the mill until it burned in 1902. From Dick Hall he bought the little store by the first wharf at the south end of the cove, which had been built in 1892 by Dick's uncle, Bob Hall, and when a high tide carried the wharf away, he persuaded the federal government to replace it with a better one. His sons built a store on the new wharf, and a cannery nearby. There was no church as yet on the island, so Pidcock continued to hold services in his home.

The story of this strict Sabbatarian ends on a disarmingly human note. He is remembered as setting forth to deliver beer to a gang of prisoners who were clearing trails on the island. This was in 1902, just before he left for Victoria, where he died the same year and was buried in Ross Bay Cemetery. The historic old Pidcock house is no longer standing. It burned to the ground in December 1977.

Pidcock's neighbour, Black Jack Bryant (so called, it was said, not for his card-playing but because of his very dark hair) left his young wife on their farm in Comox in 1888 while he went ahead to Quadra to choose property and build a log cabin. He is listed in the 1891 *Comox Directory* as "farmer, Valdez Island" but delayed pre-empting until 1911. Black Jack and settler Tom Baccus travelled to Comox in a large war canoe loaned by the Cape Mudge chief, and in this aboriginal vessel Mary Bryant and her belongings were paddled the 50 miles back to Quadra.

It was a case of survival of the fittest in those days, as there was no doctor on the island or in Campbell River for many years. On one occasion Black Jack, feeling ill, was determined to seek medical advice. He rowed across to Campbell River and then walked to the nearest doctor, who was in Comox, 40 miles south!

The Bryants went down to Wellington on Vancouver Island for the birth of their first child, William. When the baby was five weeks old they travelled back to Quadra on the steamer *Barbara Boscowitz*, which landed them by rowboat at the Cape Mudge Indian village in a howling blizzard. With their tiny infant, the young parents sheltered overnight in a logger's deserted shack on the beach, and the next morning, slipping and sliding through the deep snow, they climbed the steep bank and struggled through the forest to Dick Hall's cabin where they stayed for two weeks. When the weather cleared, the Indians paddled them to Quathiaski Cove, where they could follow a logging trail to their home. There was still no doctor available when their next child was born on Quadra. The *Colonist* of September 4, 1892 reported: "The first birth of a white child on Valdez Island occurred last week, the wife of John Bryant presenting him with a daughter."

Bryant's wife Mary, when she came to Quadra, kept a long pole in her kitchen. She planned to brace this between floor and door as a barricade if Indians besieged her cabin. Only a year before her arrival, the Lekwiltok had attacked and destroyed the schooner *Seabird*. A young Lekwiltok woman had been encouraged to come aboard. She was given whisky to drink, then locked up below and raped by members of the crew. Her husband heard her screams and came aboard demanding her release. The crew refused. The Indian and his brother tried to rescue her by smashing the hatch with axes and were shot at by the crew, whereupon other villagers joined the brothers, attacked and killed the crew, and plundered and burned the schooner.

By 1906, Mary Bryant's fears had been laid to rest. Recalling her early pioneering days in a speech at Heriot Bay, she said: "I was the first white woman to set foot upon Valdez Island, having arrived here with my husband 17 years ago from Old England. . . . One by one I have seen the settlers arriving. . . . Where, in the old days our trusty dog and rifle were our only protection from danger of animals and Indians, we now have the protection afforded by the law, and the knowledge that a neighbour is within reach of our call in case of sickness or death. The wild animals, which once we feared, have sought other homes, and the Indians are our neighbours and friends."[14]

Wolves were a constant threat in the early years, killing any domestic cats that strayed from the doorsteps and terrifying the timid with their howling. Tom Baccus had built a sturdy, two-storey log house in 1887 on the property near Quathiaski later owned by Constable Jones. One night he was surrounded by several wolves when he was walking home through the woods from Heriot Bay. Fearful of attack from the skulking shapes, he shinnied up a tall tree and spent a sleepless night in the branches. A few years later, a severe winter decimated the wolf packs and eventually they disappeared from the island.

The farms of the Joyce brothers, Alfred and Walter,

ran down to the cliffs at the south end of Cape Mudge, where they had 1300 acres between them. Farmers often supplemented their earnings by fishing or logging, and Alfred's cheerful, sturdy, Swiss wife undertook the farm work at such times, harvesting the wheat and grinding it into flour. When meat was needed, she loaded her rifle and headed for the woods to shoot a few grouse or a buck deer. Her son was the first boy born on the island, in 1896, and to mark the event his parents named him Arthur Valdes Joyce.

More than one woman on the island was an expert marksman. It was young Aileen Hughes' duty to keep the family supplied with pheasants. These were plentiful for some years, but farmers grew irate when the birds demolished green vegetables and grain; they set out poison which eventually eradicated the beautiful game-birds.

Tom Bell's 191-acre farm (later sold to Yeatman) lay between Quathiaski Cove and Gowlland Harbour and was surrounded by Pidcock property. Bell became a regular correspondent for the Department of Agriculture. In his 1895 report he says the settlers sold potatoes, pork and eggs, but marketing was difficult, the nearest outlet being 40 miles by water in a small boat and then 12 miles overland by coal train to the Comox mines. Bell also complained of the lack of roads on the island. He wrote with enthusiasm of the abundance of fish and game, and of the immense root crops in the gardens, but deplored the preponderance of bachelors who took up three-fourths of the land and were off their places half the time, adding little to the island's development.

The logging camps proved to be a good market for farmers during the early years of settlement, and the British fleet was also welcomed when its sailing ships anchored in Drew Harbour to practise manoeuvres. Its sails were usually sighted from Cape Mudge, and the alerted farmers loaded their wagons, hitched up their horses and hastened to the harbour. There was a certain amount of cattle raising on the island, but once the big logging companies pulled out, the high cost of freighting to outside areas discouraged most settlers from raising more beef than was required for their own large families.

Early settlers sometimes took up land the loggers had cleared, though this could be an isolated part of the island and meant a lonely life for their wives. The William Laws who came from New Zealand in the early 1900's chose land logged by Hiram McCormick in Hyacinthe Bay and converted McCormick's two-storey logging store and warehouse into a farmhouse where they raised three children. There were no connecting roads then from Hyacinthe Bay. To get to the Heriot Bay school or store on foot involved a long hike over a narrow trail that wound around coves and up and down steep slopes.

Tom Leask, a Scotsman who has become the legendary Paul Bunyan of Quadra Island, pre-empted 160 acres in 1889 in Hyacinthe Bay. He is the symbol of the rough, rowdy days of the early logging camps and saloons, though, oddly enough, he was not a large man and spoke in a rather soft, gentle voice. Tom was distinguished by a double row of teeth, and liked to pick up one of the very thick beer glasses used in the hotel bars and bite a sizeable chunk out of it. If sceptics declared he succeeded due to a flaw in that particular glass, he would entertain his companions by biting several glasses in a row and setting them up side by side on the counter. He is also said to have fought a belligerent logger with his bare fists in a battle that lasted throughout the night and the whole of the following day.

In the late 1880's, Tom rowed up the rugged British Columbia coast from Vancouver to the Queen Charlotte Islands, seeking shelter in the numerous bays and inlets from the fierce storms that beset those waters, and always looking for an ideal place to settle. From the Charlottes he brought back a Haida chief's daughter as his bride, after impressing the natives by rolling aside, unaided, a huge boulder that blocked the skidway they were building to launch their immense dugout canoes. He settled with her in Hyacinthe Bay, and when she died shortly after their fifth child was born, he raised the children himself, refusing to part with them to their Haida kin who came south to claim them. Despite the many dangerous miles that Tom had travelled by rowboat without mishap, he met his death by drowning, in sight of his own cabin, when his overloaded skiff capsized in Hyacinthe Bay.

These were the days of rowboat and canoe travel. After 1904, when a store was opened in Comox, settlers took turns rowing the 40 miles to that town for groceries. The S.S. *Joan* had a port of call at Cape Mudge, but all steamers found the landing too difficult in bad weather and their arrival was uncertain. In 1891, the Union Steamship Company launched the *Comox*, the first steel ship built in British Columbia, to serve the remote logging camps and settlements. She was 101 feet long, with a speed of 12 knots, and she carried passengers and a variety of freight, from mail bags to poultry and pigs. Heriot Bay was one of her ports of call.

Yet, despite the known hazards, settlers continued to row their little boats through the turbulent waters of the passage. Ernest Halliday, brother of Indian Agent William Halliday, sought shelter on Quadra Island during a storm when he was travelling by rowboat from Kingcome Inlet to Comox with his wife and two small children. The 150 miles from Kingcome to Comox made a two-week trip by rowboat. Among the many who lost their lives in the rough waters was Bob Hall, the postmaster and storekeeper at Quathiaski, who drowned in 1896 when his boat was caught in the tide rip somewhere between Cape Mudge and Seymour Narrows.

The tide in this area ran from six to 12 knots an hour at flood and from six to eight at ebb. Flood and ebb each ran for about six hours and still water lasted for only about ten minutes. Chester Haas, whose father, Raymond Haas, pre-empted land next to Walker in 1909, tells how his family was blissfully unaware of dangerous currents and travelled up to Gowlland Harbour in two rowboats lashed together, with all their belongings piled high on boards laid across the gunwales. They were met by the horrified minister who told them they had come up the passage during the brief interval of slack water; ten minutes later their top-heavy craft would certainly have been swamped in the flood.

The build-up of fog in the narrows was another problem. During the years of the gold rush, ships loaded with miners and barking dogs often entered the passage and sometimes were marooned by fog at the entrance. Settlers would row out to chat with the passengers and crew and hear their tales of fortunes made in the goldfields of Alaska and the Yukon.

-5-

The shrewd Lekwiltok elders soon recognized that the white man had come to stay and that survival depended upon learning his ways and studying his skills. Schools for Indians in pioneer times were financed by the Department of Indian Affairs and run by missionaries with varying degrees of education. The Agent's Report of 1884 mentioned that the Cape Mudge Indians were anxious for a school, and the following year Agent Blenkinsop announced that it had been decided to open a school at the "Laichkwiltach" village near Cape Mudge.[15] However, it was not until eight years later, when Billy Assu, 23 at the time, boarded the mission boat of Thomas Crosby and urgently pleaded for the school that wheels were set in motion. Crosby contacted Indian Agent Pidcock who asked his department for lumber. Methodist missionary R.J. Walker was engaged and the Indian school, the first school on Quadra, was built in 1893. For many years it served a double purpose as church and school for the Indians.

Settlers erected the first school for white children of the island in 1895 on land donated by settler John Smith near the Hughes ranch on Heriot Bay Road. Kate Smith was the first teacher at the tiny log school, and boarded with the Hood family. The School Act of 1895 required at least seven children to attend school regularly before the province would supply lumber for a school or pay a teacher's salary. Fortunately, pioneers as a rule had large families, and as soon as enough children reached school age, a building was requested.

Methodist missionary Walker, who came as teacher for the Cape Mudge Indians, had left his posting at Port Simpson only the year before to settle in Nanaimo with his wife, the former Agnes Knight, matron at the Crosby Indian Girls' Home in Port Simpson. From these postings of comparative comfort they came to the most primitive accommodations. In September of 1893, having travelled from Nanaimo on the steamer *Barbara Boscowitz*, Walker, his delicate wife, two little girls and a few household possessions were landed by rowboat on the beach in front of Yaculta, the Cape Mudge reserve. Billy Assu helped them unload their baggage and packed their bureau on his back up the beach to the small shack which was to be their first home.

This was the cabin referred to in Pidcock's diary on April 20, 1888 when he wrote: "Canoe of Nimbiish [Nimpkish] Indians arrived with letters from Mr. Hall, who told us . . . the *Glad Tidings* had taken some lumber to Cape Mudge to put up a residence for the missionary, a Mr. Reid."[16] (The *Glad Tidings* was the Methodist missionary boat.) George Reid, however, stayed only three weeks in the village. J.E. Galloway who arrived in 1892 stuck it out for nine months, just long enough to build the "residence", a 12-by-18-foot cabin.

The Walkers described the Indian village at this time as consisting of ten or more large houses, 70 to 90 feet long, with three or four families living in each, and a few smaller houses which had been built by the younger men, strung out along the shore. Tall totem poles stood before the houses of chiefs and subchiefs. The big houses had earthen floors with a fire built in the centre and a hole in the roof for smoke to escape. Each of the four corners had wooden flooring with space left for a fire, and a wide shelf against the wall for sleeping. In these corners the individual families lived. Pots stood on the floors and dishes were kept in biscuit boxes nailed to the walls, while over the fireplace the Indians kept racks on which they dried their fish.

Missionary Walker encountered a number of frustrations in his new posting. The Indians came to his services and sang hymns heartily, but afterwards drums would sound as the elders summoned the band to the Indian ceremonies Walker had just denounced. The schooling had its frustrating side as well. The children played truant, running off to hide in the woods during teaching hours; boys were taken from school to watch the potlatch dances; whole families left the village during seasonal food-gathering periods and the school was empty, though Chief Assu encouraged attendance at other times and even sat in on a few classes himself.

Agent Pidcock reported to his department in 1894: "When the Indians are at home the attendance is very good, but they do not seem able to give up their nomadic habits and consequently the average attendance is very small. The Methodist missionary, Mr. Walker, with his wife, who took up their abode on the reserve nearly a year ago, have already done good service, and have been a great check to the introduction of liquor in this tribe."[17]

In return for their school, the Indians were told to become Christians, renounce alcohol, abolish the potlatch and give up their secret dances. Agnes Walker said the Indians had expected them to "to teach school and preach, care for them while sick and help them generally but beyond that we were to let them alone. But they were soon undeceived and found they could do no wrong in peace."[18]

Two young Indian girls came to the mission house one winter night and asked Walker's help. The villagers were celebrating with a drinking party and the girls wanted their brother brought away before a fight started. Walker got his lantern and his dog and walked three miles along the beach and through heavy timber to the Indian Agent's home, getting there around midnight. The two went to the chief's house, arrested the ringleaders and clapped them into jail. Court was held the next morning, and those who had supplied whisky, as well as several others, were severely fined under the 1884 Indian Act Amendment.

Many of the Indians were angered and kept their children home from school the next day. The men sat on the school fence listening to the ringing of the bell and shouting that Walker was no teacher but a "half constable." From then on, however, Assu did his best to keep liquor out of the village, and on one occasion publicly whipped a whisky pedler who returned to the reserve after a warning to stay away.

Pidcock reported in 1894 that several Cape Mudge Indians had expressed a willingness to farm. They were the only tribe in the agency who owned livestock, having about 14 animals. The village also possessed a plough, spades, mattocks, hoes and rakes. Farming was completely alien to the west coast Indian way of life, but it was hoped that tending their own plots of land would end the migratory habits that took the children out of the schools.

However, the livestock at the Cape Mudge village was soon reduced to one cow and one horse, and not long after that the horse was shot by mistake. By 1910, Agent Halliday reported in chagrin that "very little garden stuff is raised. . . . Their ideas of the ideal and that of the whites do not at all correspond. Their chief aim is to go through life easily and get all the fun and glory they can out of it."[19]

Missionaries and agents up and down the coast were violently opposed to the potlatching and dancing. The Indian way of life, with its migratory seasonal wanderings to food-gathering areas, its lay-offs for feasts and potlatches, its emphasis on dignity and pride and the acquiring of wealth only in order to give it away, was the reverse of the white man's ideal of steady work and the saving of money.

Bombarded with complaints, Prime Minister John A. Macdonald told the House of Commons it was utterly useless to introduce orderly habits among the west coast tribes while the potlatch was in vogue. He said the Indians met and carried on a sort of mystery, remaining for weeks and sometimes months, as long as they could get food, and carrying on all sorts of orgies. That same year, 1884, the Indian Act was amended to make whisky dealers liable to punishment if they sold to Indians, while anyone giving a potlatch or encouraging an Indian to give one was liable to imprisonment for a term of not more than six nor less than two months.

Agent Pidcock complained the potlatch was impossible to control as the Indians believed they must repay their debts or suffer a lifetime of shame. Judge Matthew Baillie Begbie complicated the problem by stating that some dance potlatches were innocent, and that the potlatch was acceptable unless liquor, rioting and debauchery were involved. So the potlatches continued, especially among the defiant Kwakiutl, and the missionaries continued their complaints.

A virulent account of a Cape Mudge potlatch was published in the *Toronto Empire* of 1893 by a group of missionaries who came from Comox to Cape Mudge and found 1200 Indians from a 100-mile radius gathered there for the ceremony. The hordes of guests were crowded into temporary tents, and the missionaries deplored the resultant litter and filth, which must have resembled the aftermath of some present-day "be-ins." The sight of naked children playing amongst the refuse with the Indian dogs, some of which were licking out empty pots or lying on the beds, the missionaries called "disgusting" and a "seething mass of corruption." They were certain that morality must be at a very low ebb, and declared that "when a score of white men come in with a few gallons of 'fire water' and spend the night with the Indians, the scene becomes Indescribable . . . the Indians, instead of being an upright and industrious people, are a filthy, indolent, degraded set, a disgrace and a curse to our country."[20]

The potlatch described by the Comox missionaries may have been the one anticipated by the *Colonist* in 1892: "The Indians of Cape Mudge and Valdez Island are preparing for another big potlatch, which promises to eclipse the one recently given by Salmon River Bill in point of liberality of the gifts and the numerical strength of the gathering. Orders have already been given for thirty canoes, $1,000 worth of bracelets, 700 boxes of biscuits, 2,000 blankets, 600 barrels of flour, 400 trunks and a great variety of miscellaneous articles. The big event will commence early next week, and will last for several days. Twenty canoes are in port loading freight for the potlatch, which promises to be one of the biggest held in years."[21]

The potlatch was a deeply-rooted custom among all the coast tribes, a chance to prove a man's superiority, and not necessarily impoverishing, as a gift of equal or greater value was expected in return. Since potlatches were also held to mark birth, coming of age, marriage and death, they served as public announcements confirming these important milestones, vital for a people who had no written records.

Chief Billy Assu has said: "My grandfather was still alive when I was a youngster and in my teens he told me the history of my people. I learned who was who, who were the chiefs and their descendants, who were strong and powerful, and who were weak and amounted to nothing... before a man is permitted to relate or make history he must first prove his worth by deeds and by valour, and by giving many great and expensive potlatches."[22]

Assu had an outstanding record as a potlatcher. The Indian names he acquired during potlatches at his birth and his arrival at adolescence, and the names he took himself at potlatches that he gave after maturity, all included a phrase meaning "giving away." His potlatches were numbered in the hundreds. At the time he became chief he entertained over 2,000 guests from 23 tribes for two weeks and gave away $25,000 in gifts and food. He built himself a Chief's Big House and had two houseposts carved by Johnny Kla-wat-chi of Alert Bay. For the opening of the Big House, Assu invited all the chiefs of the Southern Kwakiutl nation to his potlatch, and brought a famous speechmaker from Fort Rupert to describe the greatness of the Wiwekae and their chief.

Not all the ceremonies were secret. Visiting explorers were welcomed with a dance that scattered swansdown over them, signifying peace. Other dances were purely for entertainment, full of sly humour or slapstick comedy, and white men were sometimes invited to attend. One held around 1890 at Cape Mudge was des-

cribed by Michael Manson, trader and Justice of the
Peace from Cortes Island, who was on friendly terms
with the Cape Mudge Indians.

Manson was a guest of Chief Quocksister, who seated
him in a place of honour, wrapped in a blanket covered
with pearl buttons and wearing a head-dress of bear
claws. During the show, the actors came out dressed like
ducks with long bills and with wings attached to their
arms. They quacked, waved their arms, went off the
platform, returned again, quacked, waved, departed,
and continued this routine for over an hour. The producer explained they were telling the people they had
found good feeding grounds. When the repetition
appeared never-ending, the duck bills were suddenly
knocked off and the faces of the Indians, which were
greased, were slapped with bags of cheese-cloth filled
with flour. "Now they have become white men,"
announced the producer, and the Indians "almost lifted
off the roof with glee."[23]

Manson said the producer of the fun-making skit was
"one Chekite", who may have been Hereditary Chief
Johnny Chickite who drowned off Cape Mudge shortly
after this, at a time when his son was still a child. Assu,
who early displayed a talent for leadership, was therefore
appointed chief. Physically, Assu met the old-time tests of
superiority: he possessed strength and endurance and was
a highly successful hunter and fisherman. "Also," his
granddaughter Audrey Wilson said, "he had great skill in
speaking... he could hold attention and move us to feel
as he did. He never asked; he told us what to do."[24]

Later, Assu married his daughter Lucy to Chickite's
son. Young Johnny Chickite became hereditary head of
Hamatsa, and custodian of the clan history in the days of
the big houses. He knew all the potlatch feast songs, the
cedar-bark songs, the dancing done during potlatch
feasts, the giving-away-wealth songs and the mourning

songs. All were forgotten, Chickite said, before he reached his old age.

Missionary Walker's wife, when she was Agnes Knight of Port Simpson, had heard tales of the Indians' *Hamatsa* ceremony, but was told the details were too horrible for a woman to hear. Certain winter dances which the Lekwiltok allowed her to see, however, she called "graceful and harmless."

The *Hamatsa* ceremony was known to white men as the Cannibal Dance. A young man was taken into the woods, where he was filled with so strong a spirit that he returned in a frenzy and could only be calmed if allowed to devour a human corpse. Subdued by this, and by the songs and dances of the group, he danced quietly, but might be siezed again by the spell and then would run about the room biting bits of flesh from the bodies of the spectators.

In the distant past the dead were placed high in the branches of trees where the bodies mummified, and for the *Hamatsa* a corpse was taken down and soaked in salt water, its extremities removed and the decayed flesh scraped from the skin. It was then taken to a hut in the woods where the *Hamatsa* split it open and smoked it over an open fire before using it in the ceremony. Alice Joyce of Cape Mudge, daughter of a Gilford Island Kwakiutl chief, said: "In the old days they did things I'd hate to tell you; it was cannibalism, but that was very long ago. My uncle is *Hamatsa*. Nothing like that has happened for a long time."[25]

Curtis[26] tells of the initiation of a 12-year-old Cape Mudge boy into the *Hamatsa* society during the winter of 1909. One of the *Hamatsa* members was instructed by the chief to collide with the boy while he was running about the village in the frenzy caused by the spirit. The boy was told he must then fall and lie motionless as if dead. When the boy collapsed, word was spread that he had been killed by the *Hamatsa*; the father searched the village with a gun, demanding revenge, and the women screamed and scratched their faces in mourning. The chief told the people the *Hamatsa* had taken the boy's life and the body must be buried in the woods. As soon as the boy was carried away, whistles were heard in the forest, a sign that he would become a *Hamatsa*. Winter nights were cold, so the boy was brought back secretly to his house and kept there until it was time for him to enter the ceremonial house through the roof and perform his dance, naked except for bracelets and a crown of cedar boughs.

The *Hamatsa* was looked upon by the whites as debauchery. Early missionaries, traders and explorers who saw parts of it, or heard tales of its horrors, expressed their loathing and disgust. Much of the horrifying aspect may have been the result of sleight-of-hand, however, as the Southern Kwakiutl were past masters of stagecraft. Ninety-year-old Louise Hovell of the Cape Mudge Indian band recollected: "We called the potlatch '*passapa*'. The missionaries thought *passapa* was our religion; they didn't understand. They thought *Hamatsa* was wicked; it wasn't wicked; it was exciting when the whistles began to sound in the woods, and it scared us when the 'wild man' came out and danced, but it was like a play. It's all gone now, and forgotten. I remember only the singing, the wonderful singing in the Big House."[27]

-6-

Life gradually became easier for the pioneers in the years between 1900 and 1920. Logging flourished, mining started up, and a cannery and sawmill were built, all providing work and encouraging more settlers to come to the island. New and better schools were supplied for a number of districts. The Columbia Coast Mission was established with headquarters at Rock Bay from which it provided a doctor and floating hospital on call for emergencies.

At the start of the century there were still hardships for settlers to endure. The children of missionary Walker at Cape Mudge started their walk to school before daybreak, six months of the year. Carrying a lantern, they climbed the almost perpendicular path up the cliff from the Indian village to the trail which led to the main wagon road. Then came a frightening, five-mile walk through the dark woods to the log-cabin school. The journey was shortened in 1903 when Walker bought Elof Johnson's 240 acres on Gowlland Harbour and the family moved to their new home, two miles closer to the school.

The first, rough, log-cabin school of the settlers was soon in a sad state of disrepair. An indignant Cape Mudge resident wrote to the *Colonist* on March 2, 1905: " . . . At the Christmas closing exercises, a gentleman from Texada Island visited our school, and he remarked that the school was not a fit place in which to house children; that he would not send any of his children to such a place, and that he could not get warm for some time. If he had said that he could not think of housing his pigs in it he would have hit the mark! I'll invite the readers of your paper to come and see the little children huddled up around the stove trying to keep warm, and on rainy days, snowy days, get up in a corner to keep their books from getting wet. It is really lovely to sit in that old draughty hovel . . . "[28]

This dramatic plea resulted in an $800 allowance for a small frame building the following year. The Pidcock brothers donated land for it on top of a hill beside a swamp, about a mile from Quathiaski Cove. Heriot Bay was made a separate school district in 1910 and given its own school, for which hotel-keeper Hosea Bull donated land, a local sawmill supplied lumber at cost, and labour was voluntary. The 22-by-24-foot frame building was in use for nearly half a century.

Settlers in remote areas were still expected to build their own schools in 1912, but volunteers were always ready to help. That year, Finnish settlers Alfred and Emil Luoma and Arthur Stenfors raised the first log-cabin school in Granite Bay on land donated by Stenfors. Hastings Sawmill Company men cut the logs and shakes and hauled them to the site with their horses and oxen. The school opened with 12 pupils and Miss B. V. Cousins as teacher. It was a constant worry that a term would begin without the required number of pupils, and a teacher with a large family to swell the enrolment was the popular choice.

Granite Bay was known as the Finnish settlement due to the number of Finns who came to work in the logging camp and stayed on to become permanent residents. Alfred and Emil Luoma, who pre-empted land on Quadra in 1903 and 1906, were born in Heikijoki, Finland, and moved to Quadra from the Finnish settlement of Sointula on Malcolm Island.

The first Finnish settler to bring his family to Quadra Island was Konstantin Wilhelm Stenfors, who came from Pori on the Gulf of Bothnia. Like many other immigrants at this time, he had hoped to make his fortune in the goldfields of the Yukon. In Nanaimo, he tried to find work as a machinist to raise money for a grubstake, but his English was limited and he had no luck. Finns working in Nanaimo told him of the northern islands and he opted for Quadra, pre-empting land in 1903 at Granite Bay and turning his hand to logging and farming.

The fear of forest fires shadowed the Stenfors' first year. A bush fire came perilously close to their cabin and a huge stump near their door burned continuously all summer long. One of the chores of ten-year-old Irene was to carry out buckets of water every day to wet down the stump and keep the fire from spreading. The Stenfors' fifth child was born two years after the move to Granite

Bay. No doctor was available, so Stenfors and his little daughter Irene acted as midwives. In later years, Irene married a Nova Scotian, William Stramberg, who was working as a blacksmith for the Hastings outfit. The Strambergs lived in Granite Bay for more than 40 years, Stramberg becoming part owner of the Geiler group of mines, later taken over by the Noble brothers.[29] Stramberg Lake, close to the Geiler claims, is named after him.

Joseph Dick, son of Mrs. Hosea Bull of the Heriot Bay hotel, opened the Granite Bay hotel in 1910. Henry Twidle, who photographed many of Quadra's early activities, took it over in 1911.

The urgent need for hospital and doctor service was greatly eased by the Columbia Coast Mission's hospital ships, which travelled up and down the coast serving logging camps and settlers in the more remote areas between Vancouver and Alert Bay. The Columbia Coast Mission had started its work in 1905 with the launching of the missionary steamer *Columbia* at Vancouver. She was 64 feet long with a 14-foot beam, and was driven by a Union gasoline engine of 20 h.p.; she was also fitted with masts and sails for use when the wind was favourable or in case of a breakdown.

The year before, John Antle had left his post as rector of Holy Trinity in Vancouver, and set out to survey the coast as far as Alert Bay, studying the needs of settlers and loggers. With only his nine-year-old son Vic as crew, he travelled in a tiny, open sailboat, the *Lavrock*, which he had fitted with a ¾-h.p. Springfield "Bull Pup" gas engine. The engine, a novelty on the coast at that time, mystified John Davidson, the lighthouse keeper at Cape Mudge, who saw the boat when it first rounded the cape. Antle wrote in his memoirs: "The lightkeeper, a Scotsman named Davidson, said afterwards: 'I mind the day when ye kem roond the Cape. Y'r sail was up but it was no pullin' and yet ye gaed along at a guid rate. I was fair surprised till someone tell't me aboot the contraption called the gasoline engine, and then I kenn'd a' aboot ye.'"[30]

A glimpse of the dangers and discomforts endured by the two voyagers in their little craft, which resembled a rowboat fitted with a sail, is given in Antle's account of their homeward trip through the Yaculta Rapids. Antle was preparing to start his engine after a night spent in a logging camp when the foreman came down and invited the travellers to breakfast with the crew before setting out. Antle had chosen the early hour in order to hit the rapids at slack tide, but by the time they had shared the loggers' hearty breakfast, the rapids were at full flood, running very strongly. The two-mile course was crooked and they were unable to gauge the extent of the dangers ahead until they neared Dent Island and saw whirls and turmoil beyond it. Antle advised his son to inflate his air cushion and hang on to it if anything happened, which "had the effect of thoroughly scaring him."

He continues: "We skimmed along the edge of several big ones, looking with dread into their funnel-like depths, with the question in our minds: How long before we shall take a header into one? Sure enough, almost immediately we found our little vessel at an angle of 30 or 40 degrees, skimming around the edge of a big one. Finally, we crashed to the bottom of it with a bang against a chunk of wood, which shook every timber in the little ship. The boy was yelling at the top of his voice and grabbed his cushion, while I, less articulate, but just as scared, wondered how far down we should go. But to our great surprise and relief the smash into the bottom of the pool destroyed its whirling motion and we were suddenly on a flat surface, the engine and sail both still doing business."[31]

Antle was appointed skipper of the *Columbia*, which carried a medical and surgical bed and a full equipment of appliances. Dr. Hutton was ship's doctor. The Columbia Coast Mission proved a godsend for men injured in the camps. Besides offering energy treatment on board its ships, the mission built and supported (with the help of the government and of logging, mining and fish-packing companies) hospitals at Rock Bay, Vananda and Alert Bay, to which badly injured men could be rushed, avoiding the long trip to Vancouver or Victoria by steamer. One who benefitted was Arthur Stenfors, the 18-year-old son of Konstantin Stenfors in Granite Bay, who was working as brakeman on the logging train and was terribly crushed when he was caught between the car and the skids of the landing. No one expected him to survive, but he was rushed at once to the Queen's Hospital at Rock Bay and to everyone's amazement he returned some time later "practically as well as ever."

Not only loggers were served. Antle records in 1907 that Dr. Beech of the mission hiked the miles from Gowlland Harbour to Heriot Bay to see a man ill with measles. On another occasion, a year earlier, the mission boat was at Granite Bay visiting Paterson's camp when the launch *Thistle* arrived with a message from Constable Jones of Quathiaski requesting a doctor for his child who was dangerously ill. It meant a long haul by sea, travelling north to Rock Bay to pick up the doctor before returning through Seymour Narrows to Quathiaski. From there, a horse and buggy took the doctor three miles inland to his patient.

"Little Jimmy", a portable organ, was carried on the *Columbia*, and *Rendezvous* for over 70 years, first by the Reverend John Antle and later by Canon Alan Greene, on the ship and in the bush, for services, christenings, weddings, funerals or singsongs. Old-timers remember Alan Greene setting it up on the beach to play to Indians in front of the reserves. The organ, and the bowl and compass from the first *Columbia* (later transferred to *Columbia II*) may be seen in Vancouver's Maritime Museum.

Alan Greene joined the mission in 1911, the Reverend C.C. Owen having arranged the use of the little Howe Sound boat *Eirene* for him. Unlike Antle, who had sailed the wild coastal waters of Newfoundland for seven years, Greene arrived fresh from the University of Toronto, with no experience in navigation. He was promptly made skipper and engineer of the *Eirene*, a 30-footer with a balky, two cylinder ten-h.p. Palmer gas engine, though he admitted that valves, carburetors, coils, pumps, clutches and gears were just words to him.

Antle's practical method of bringing religion to the loggers and settlers was to emphasize the mission's services of doctor, hospital and library. (To offer only religious services, he said, was "putting the thick edge of the wedge first.") Greene, with no "thin edge" on his little boat to attract possible converts, found his first years hard going. He tells in the *Log*, the Columbia Mission paper, of an entire congregation at Heriot Bay consisting of an agent for Disston saws; minister and congregation climbed through a window of the schoolhouse and "joined in corporate worship."

There were ministers in the busy days of the logging camps who tried every imaginable dodge to get the loggers to listen to their sermons. A few are said to have held their services in the beer parlours, the one spot where they could be sure of an audience. George Parsons, a Quadra pioneer who settled near Waiatt Bay at the turn of the century, said he saw blackjack games and sermons going on in the same building. One minister used to pick up cards and bottles at Shoal Bay for loggers who had ordered them, and when he delivered them would give a quick sermon on the evils of gambling and drinking as he handed them over in the homes of each of the men.

The first church on Quadra was St. John's, built in

1917, with the minister, Mr. Comley, doing much of the construction work. Mrs. Smith, mother-in-law of Mary Pidcock Smith, donated a portion of her garden for the site, on a hill overlooking Quathiaski Cove. The missionaries said the meaning of Quathiaski was "Peace Pool," which they thought appropriate; however the small island in the cove was once a fortified Salish village and the Indians give the meaning of the original Salish word, *Qatasaken*, distorted by whites into Quathiaski, as "a mouth with a bite of something in it", referring to the cove with an island in it. The building was dedicated by the Right Reverend Bishop Schofield, and Alan Greene, rejoining the mission after wartime service overseas, was appointed rector. For 17 years the church flourished, with missionary Walker's second wife, the former Quadra Island school teacher Theodora Spencer, active in its Women's Auxiliary.

A colourful couple who feature largely in the island's development were Hosea Bull and his second wife, Helen. Bull bought up most of the land along the Heriot Bay waterfront. In 1901, when he subdivided and sold a part of his property, he donated a narrow strip along the waterfront as a public esplanade. He built the first Heriot Bay hotel and saloon, with Charles Hodack as part owner. Old-timers say this was in 1894, though "Hotel Heriot" is first mentioned in the 1901 *British Columbia Directory*. To encourage loggers from the Gowlland Harbour camp to come to the hotel saloon, Bull cut a wide trail through the bush from Gowlland Harbour to Heriot Bay. Here the men were greeted by the bar with its brass rail, and spittoons for the devotees of tobacco and snoose. Only the most obstreperous were evicted to settle their differences on the grounds behind the hotel.

Bull operated a sawmill on the waterfront, on the site of the present Heriot Bay store, until it burned down a few years later. He ran a logging camp, and continued his early trade as baker by supplying his camp with bread, as well as selling it to the ships of the British fleet which entered the nearby waters on manoeuvres. He also ran the Heriot Bay hotel, store and post office. The living room of a house that he built for his relatives, the Dicks, on the waterfront on the present Mowat property became the post office in the 1920's and '30's. The old house was recently demolished.

The Heriot Bay hotel (now Heriot Bay Inn) was built, burned, rebuilt and changed its ownership so many times that its history is difficult to trace. It was frequently altered; at one time it was in two parts with a walk between. The first hotel was built farther west along the bay from the site of the present inn; later ones were on the site of the present building by the huge maples near the wharf. One of the hotels opened December 11, 1911 and burned down on May 11, 1912. George Pidcock recorded the event at the time in his diary. There was no fire department on the island in those days and once the flames took hold it meant a total loss. Only the hotel chimney was standing when the fire burned itself out.

Under the second Mrs. Bull, a New Zealander, the new hotel that replaced it took on an unfamiliar character of respectability and social nicety. The beer parlour still flourished, offering relaxation to weary loggers and fishermen, but a gentler clientele was served by the remainder of the elegant rooms. There were 19 bedrooms and an upstairs dance hall, as well as a wide verandah on the second floor, used for summer dances. In this era, hanging flower baskets decorated the lower verandah, which served as a tea-room, and young girls clad in white uniforms waited on the guests. There was an aviary of rare canaries, and old-timers say a pet seal was kept in a pool behind the hotel. Helen Bull was an attraction in herself; she was a tall, handsome woman with snowy white hair, and was seldom seen without her equally snowy-haired little Pomeranian dog.

The schedules of the coastal steamers were always uncertain, and travellers were expected to wait several
hours at the wharf in the dark and cold to avoid missing
their vessel's brief docking. They were invited by the
hospitable Mrs. Bull to use her warm, comfortable parlour as a waiting room. She gave lavish entertainments,
one of which was described by Mrs. B.E. Ward, lumberman's wife, in the *Log*, in 1906. Five full pages long, the
report reveals a determined attempt to soften the rough
edges of a farming, logging and mining community on a
remote northern island.

Mrs. Ward wrote in part: "The most interesting event
that has ever taken place among the B.C. Islands was
witnessed at Heriot Bay at high noon, on the 3rd of July.
For the past month, invitations have been sent out to the
residents of Valdez Island, adjoining islands and camps.
The principal feature of the celebration was the christen-
ing of the new steam launch, which is to be used to convey visitors of that noted hunting and summer resort to
the various points of interest among the B.C.
Islands . . . there were over three hundred guests. . . . Long
banquet tables, draped in snowy linen, sparkling with sil-
ver and crystal, and garlanded with silken flags and cut
flowers, were spread beneath the luxuriant maple trees
upon the lawn before the Hotel. . . . Master Cecil Bull (the
eight year old son of Mr. Bull) clothed in a spotless suit of
white duck, took his stand upon the upper deck of the
launch, and gracefully invited all guests who could find
standing room upon the launch to come on board."

After Mrs. Bull had christened the launch with cham-
pagne, three ladies, "each robed in dainty clinging gowns
of sheer white lawn", stood in the bow of the launch and
recited poetry at some length. A portion of one stanza
was this:

> "Oh NEPTUNE, thou god of the restless deep,
>   We give this treasure to thy arms,
> Let her upon thy bosom sleep

Safe from all alarms;
With kisses, bathe her virgin breast,
Clasp her in close embrace,
Safely may she ever rest
In that beloved place..."

One wonders how a crowd of ribald loggers reacted to these lines ("Frank Gagne's camp was in attendance in full force") but Mrs. Ward says "Each one present was on their very best behavior" and once the recitations were at an end they were able to let off steam: "Cheer after cheer rang out from every throat, then was repeated three times three, until the crowd was hoarse with shouting."[32]

Telephone service came to Quadra early in the century. George Pidcock in his diary tells of a telephone cable being laid on November 15, 1910, and on November 24 says he "spoke over the phone for the first time to Comox." Trevor Bagot was the first lineman on Quadra. His family came to Quadra in 1907 and bought the property of Constable Jones in Quathiaski Cove opposite Mary Island. Trevor's brother Seymour, who became Quadra's Justice of the Peace in 1912, says the phone cable was laid from the mainland via Cortes and Mary (Marina) Island to the beach by Bagot's farm. The phone station there was called "Bagot's".[33] From Bagot's the wire ran to Quathiaski, and by submarine cable to Grouse Island, overland across Grouse, and underwater again to Campbell River. This cable was later replaced by one laid underwater the whole way, submerging on the Quathiaski side south of the old ferry landing.

Ships heading up the channel would drop anchor in mid-stream to wait for slack tide before attempting Seymour Narrows: the strong tide would cause the ship to drift back, and the anchor would drag and catch in the cable. Captain Tom Hall often went out to caution ships about the cable. If they were caught, they sometimes slid over to the Campbell River side where the slack in the cable allowed them to free their anchor. Lineman Eddie Joyce told of one ship that caught her anchor, whose captain wanted to chop the cable to get free. He was informed he would be charged $5 a foot for a new cable, whereupon he chopped off his anchor instead and sailed away without it.

Telephones in the early days were all party lines, and callers cranked a handle to ring four short and two long, or five long and a short, or whatever number was wanted. The rings sounded in every house with a telephone on the line. Some callers cranked out tremendously long rings to be sure of distinguishing them from shorts, and when any such muscular arm rang six longs at five in the morning or twelve o'clock at night, a large part of the island population rolled over in bed and cursed the caller. "Listening in" was a favourite pastime. This interfered with the reception of long-distance calls, but after a request to anonymous listeners to "hang up, please, this is long distance", a series of obedient clicks announced the clearing of the line.

In the earliest days of white settlement there was no cemetery on the island and burials took place in any convenient location. The point at the eastern end of Heriot Bay was occasionally used, which explained the skeleton that Rowley Grafton uncovered when he bought the property and was digging a trench for a gas line.

Frank Gagne, a logger from Gaspé, Québec, who married pioneer Black Jack Bryant's daughter Daisy, agitated for some time about the need for an official cemetery on the island, and land for it was finally set aside in 1913 on Heriot Bay Road near Quathiaski. It is like stepping into the past to decipher the names showing faintly on the old tombstones that lean drunkenly in the unkempt graveyard, overgrown with daisies and ferns:

names that include Bryant, Luoma, Yeatman, Pidcock, Walker, Joyce, Callow, Leask, Wilson, Law and Hall. There are unmarked, moss-covered mounds that are the graves of victims of the 1918 influenza epidemic, brought to Quadra from surrounding islands which had no cemeteries at the time. Some at the rear of the cemetery, near the swamp, show signs of excavation. These were graves of Chinese, whose relatives disinterred the coffins and sent them to China once they had raised the transportation fare.

The earliest dates on the tombstones are those of Agnes Knight Walker and Alice Pidcock, both of whom died in May of 1915, but Quadra Island calls its first grave that of 16-year-old Alice Bryant. The dates on her gravestone read 1897-1915, but George Pidcock's journal records on June 17, 1913: "Alice Bryant passed away. Funeral 19th." The cemetery was not yet prepared, however, so Alice was buried on the Bryant homestead and her coffin removed to the graveyard in 1915.

Three old gravestones, side by side, commemorate the deaths of pioneer Frederick Charles Yeatman who came to Quadra in 1894 and two of his sons, Sam and Fred, all of whom died in tragic accidents. Fred Yeatman Senior was lost while he was on a hunting trip on Vancouver Island with his son Tom, and a joint headstone in memory of Fred and his wife rests above her single grave. The *Colonist* of November 20, 1903, said Tom was found by two men on Vancouver Island above the narrows where he had been waiting two days beside his boat for his father to rejoin him. George Pidcock's launch was reported to be out looking for Yeatman, and Hosea Bull had organized a search party, since "in these wild regions no man's life is safe against the elements."

On December 16, the paper reported that Bryant, Joyce, Hughes and Russell had returned from an unsuccessful search. They had found Yeatman's rifle on a river bank and concluded he had fallen into the water and drowned. Pidcock's steamboat was out of commission (George Pidcock refers to it in 1903 as "little Stinker" because of its frequent breakdowns) so the searchers had been forced to cross the narrows in a rowboat in dense fog. At night they had heard the steamer hooting to feel its way by echoes from the steep cliffs on either side of the passage.

Fred's two sons, Sam and Fred, were drowned at ages 27 and 29 in Heriot Bay in 1922. On calm days, this bay resembles a placid lake ringed with green islands, but when sudden southeasters blow up, even sturdy fishboats run for shelter.

The modest headstone of James Weir marks the resting place of an adventurer of gold rush days. Jumping ship from the *Royal Arthur*, a British warship docked at Victoria, Jock headed for Alaska. He worked below ground in a placer mine where novelist Rex Beach acted as winch man, and then spent 30 years trapping and goldmining in wilderness areas beside the Yukon, Novakaket and Innoko Rivers. He retired to the island where his old friend of Alaskan days, Edgar Crompton, was postmaster.

Among the weeds and daisies there is a neat tombstone inscribed "Countess C. De Almeida, Portugal, Aug. 7, 1937. Age 89 years." This marks the grave of the mother of the late Mrs. Sam Boond, widow of the Fisheries Officer. Well-spoken and talented, Maria Boond was considered an eccentric by the islanders. She spoke of a glamorous past as a model in Paris, and produced creditable paintings in watercolours and oils. At her mother's funeral she startled mourners by tightly cramming the many floral wreaths into the coffin, and thereafter always referred to her mother as though she were still alive, scolding her as responsible if a door slammed in the wind and waving cars to a halt to "give my mother a ride up the hill." Mrs. Boond is the subject of an unresolved mystery. She and a friend were found dead, locked into a room in her house overlooking Quathiaski harbour, cater-corner from the old jail house. An autopsy showed traces in the livers of a poison obtainable from painting equipment. *Discovery Passage* quoted the diary of the late George Rose, road foreman of Quadra: "January 23, 1952: The Cove is all agog with the news that Mrs. Boond and a chap named Rocky were found dead in her house last night. Presumably a poison case. It is a big shock to everyone and pretty much of a mystery."[34]

Dr. Howard Jamieson, the first doctor to practise at the Campbell River hospital (built in 1914) died at the early age of 35 and was buried on Quadra in 1916. Among the newer graves is a group of four: John and Marie Blenkin and their young daughters, killed in a head-on car collision. Residents started a campaign to raise funds for a park in their memory. This is the present 40-acre Blenkin Memorial Park on Quadra, established in 1962 for the recreational use of local residents.

The Union Steamship Company increased its service to the islands early in the century. In 1901 the *Cassiar*, the "loggers' palace", was launched and christened, scheduled at first to serve northern logging towns. A luxury ship, she supplied the loggers with a spacious saloon and smoking room. The *Colonist* of 1903 mentions both the *Cassiar* and the *Comox* as calling weekly at Heriot Bay, and Pidcock speaks of the *Coquitlam* coming into Quathiaski in 1902 with freight and loading up with lumber. In December, George Pidcock records that the *Coquitlam* "arrived at 11:30 on fire, got fire out at 2 a.m." In 1908 he speaks of the *Venture*, a ship of the Boscowitz Steamship Company, leaving for Vancouver. The *Venture* was destroyed by fire shortly after this at Inverness Cannery and her replacement was acquired by the Union Steamship Company.

The beautiful *Chelohsin* was a familiar sight to Quadra Island settlers over the years. Built in Dublin Dockyard, she was delivered to the Union Steamship Company on December 29, 1911. She had 66 first-class berths and was finished inside with fine panelling. She was also a clean ship, with oil-burning equipment instead of coal. On November 6, 1949, she was stranded on Siwash Rock near the entrance to Vancouver Harbour. It was decided that refloating was impossible, the passengers and crew were removed, and she was sold to Victor David, a fish processer, for $1500. To the chagrin of maritime interests, he and a five-man crew quickly refloated the "unrefloatable" *Chelohsin*, towed her to a North Vancouver moorage, and then sold her for $25,000 to ship scrappers from San Francisco who dismantled her there.

1911 saw the dramatic shipwreck of the steamer *Cottage City* on January 26, on Willow Point Reef, across Discovery Passage from Cape Mudge. The Indians called the spot Kahushian Point ("Place of Bad Waters"). There was no lighthouse at Willow Point and the ship lost her bearings in a fierce snowstorm. John Davidson, lighthouse keeper at Cape Mudge, worked his hand-pumped fog horn diligently, but the warning sounds from across the water were lost in the blast of the storm and the *Cottage City* foundered on the reef. White settlers and Indians from both sides of the passage quickly surrounded the wreck as soon as she was abandoned, and her cargo and fittings vanished into local hands. On board there was a large consignment of wine and whisky, which the Cape Mudge Indians appropriated, resulting in an orgy of drunkenness. Chief Billy Assu, who had begun his campaign to keep whisky out of his village, called a council, created a police force, made arrests and levied large fines. The liquor disappeared, and the village reluctantly returned to sobriety.

For their trips to Vancouver Island, and to small

adjacent islands, settlers had depended upon dugout canoes, rowboats, sailboats, and later the unreliable little two-cylinder gasboats. Now, around 1914, the first water taxis began to serve the islanders. This unscheduled service was supplied by a succession of jitney launches running from Quathiaski Cove to Campbell River. Few people travelled the route in the early days, and boat owners were loath to wait about for passengers if another job turned up. Once their passengers were landed at "the River," if a chance arose to run a travelling salesman or timber cruiser on a charter trip up the coast, the jitney would take off, leaving the Quadra Islanders stranded.

Early jitneys, according to Captains Tom and Bill Hall, included the water taxi of Harry Poole of Campbell River who used his 16-foot rowboat with a 3½-h.p., two-cycle Grey engine to run miners from Campbell River to the north end of Gowlland Harbour around 1914; a man named Sykes who ran miners over the same route in a 35-foot boat with a full cabin, and Andy Castle, whose boat with cabin was available for passenger service.

-7-

Industries developed and flourished during the first two decades of the century. Lumbering on Quadra was well established by 1900 and continued to expand. The Pidcock brothers were running their sawmill at the start of the century; it burned down in 1902 but the brothers continued to log their own land and George acted as superintendent of a sawmill for 13 years. Hiram McCormick's company was logging in Hyacinthe Bay in 1901 and Sol Reamy was hauling logs the same year on Quadra with the famous engine "Curly".

This little locomotive had an eventful history. She was an old, saddle-type Number Three locomotive, built by Marshuetts and Cantrell of San Francisco in 1869 for a contractor who was building the sea wall in that city. She was also used in the building of the Panama Canal, and later brought to British Columbia in 1881 by a contractor, Anderdonk, to build the Canadian Pacific Railway from Yale to Emory, British Columbia. His firm of contractors gave the locomotive the name "Emory". Later, she picked up the name "Curly" from a machinist who was nearly run over by her. In a moment of mingled rage and fright he swore at her, forgetting her exact name as he yelled "Damn that Curly!" The new, unofficial name stuck to her for the rest of her history.[35]

After the C.P.R. contract was completed, the Hastings Sawmill Company bought Curly, and some maintain she was the first logging locomotive used in British Columbia. Her first logging job was at Mud Bay, after which she was sent to Rock Bay on Vancouver Island and finally to Granite Bay on Quadra Island where she was used for logging and mining by Hastings Sawmill. Here she remained until she was retired in 1926, when she was taken to the C.P.R. workshop in Vancouver and, repaired and repainted, was presented to the city of Vancouver. She was kept on exhibition at Hastings Park for nearly 50 years; she was then moved to Heritage Village in Burnaby, where she stands today.

Logging railroads were a romantic part of Quadra's past. The Abbot Timber Company built one that ran from Francisco Point, at the south end of Quadra, around to Drew Harbour. A logging railroad also operated at Open Bay. Until a few years ago, some of the old railroad lines could be traced by the occasional rotting tie not yet destroyed by forest vegetation.

Alan Greene said he made his first call at the Hastings camp at Granite Bay in 1911: "Many of the loggers were New Brunswickers and Nova Scotians, with Scandinavians just beginning to come in goodly numbers. . . . The Hastings Mill camps kept their employees on into old age and became known as 'Old Men's Homes'. Hundreds of almost wornout loggers found a home in these camps and were given lighter jobs. There were no old-age pensions or government assistance grants then and Hastings saved many a penniless, homeless logger."[36]

Conditions in the early logging camps were far from luxurious. Henry Leask, son of the famous Tom Leask, tells how his father worked in logging camps where three-tier bunks had bedsprings consisting of boards and mattresses of hemlock boughs. Many loggers only worked long enough to earn a grubstake, for wealth from the goldfields was the prevailing dream in those gold rush days.

Logger Archie Walker worked during the years of World War One for the Wilson Brady Company near Quathiaski, after that company took over the Abbot Timber Company, and remembers straw bunkhouse mattresses that were infested with bedbugs. The loggers were great gamblers and would sit around beach fires at Quathiaski where the Indians had summer homes near the cannery, gambling with the natives by the light of a lantern. In camp, they played poker and black jack in the evenings; Archie had only his last pay cheque to take home when the camp closed. There were two locomotives at the camp, and when news came that the war had ended, the locomotives steamed up and down the track with their whistles blowing full blast. Archie Walker headed for town with his one pay cheque to celebrate the armistice.

The combination of loggers, Indians and liquor was still causing problems. John Antle tells of trouble at Granite Bay, where the Hastings Sawmill Company had a 100-man camp. The main camp was three or four miles inland, but there was a boom camp at tide-water. Across the creek from this was an Indian summer camp, and the boom men complained to Antle of frequent shrieks and yells that disturbed them when Indian husbands beat the young wives who tried to hold back part of the money they had earned in the camp through prostitution. The husbands were outraged, as the money was earmarked for potlatch purposes. Antle found three young girls in the logging camp and threatened to take them to DeBeck, the Indian Agent, unless they left. Next he went to the Indian camp and told the chief he must move camp at once. The chief protested they were only there in their hereditary berrying area for the berry-picking season, but the next morning they packed up and departed. Antle admits his handling of this incident caused great controversy among the loggers, Indians and squaw men in the district.

Logger Frank Gagne also complained in 1907 that Cape Mudge Indian girls had been coming to the camp to sell whisky. Girls and whisky, Gagne said, were a bad combination and slowed down the work. One of the girls, known to the loggers as "Soreneck Annie", had refused his order to leave the camp, so Gagne called in Mr. Walker, then a Justice of the Peace, who fined her $50 and costs for being drunk and disorderly. Annie was furious and berated Walker with a pungent vocabulary, ending up: "If there is any other name in English that I don't know, and is bad, that I have not called you, you are it."[37]

In November of 1903, a party of American mining men set out for Quadra Island to stake mineral claims. Travelling in an Indian canoe, they tried to cross below Seymour Narrows in rough weather and almost drowned when their mast snapped in two and their sail collapsed on top of them. Fortunately, they managed to paddle to Gowlland Harbour and safety.[38] They were only a few of the many men who sought fortunes on Quadra that year. Several mines were at the testing stage: a copper mine four miles from Gowlland Harbour and another two miles from Heriot Bay.

It was in this year that Quadra's most famous mine, the

Lucky Jim, was discovered midway between Stramberg Lake and Granite Bay. Hearsay has it that loggers who used the little train Curly first found signs of the ore. Lucky Jim was a group of four claims owned originally by George D. Mumford of New York, who also owned a group of five claims, the Great Granite group, about a mile farther north. Both groups were later taken over by the Great Granite Development Syndicate Ltd., in which Eric Hamber, one-time Lieutenant-Governor of British Columbia, was a shareholder.

The fascination that Lucky Jim holds for the visitor to the island lies in the fact that for many years the huge old boiler and the steam engine wheel were left standing in the deserted area. The boiler has since been removed but the huge wheel of the steam engine, 12 feet in diameter, stands upright in the silence of the forest, an alien object among the trees and ferns, like a ghost from the past. Some distance up the road, in among the trees, are the crumbling remains of the bunkhouses, their massive logs dripping with moss. Most of the smaller rafter logs have fallen from place, leaving the structures open to the sky. Very soon the forest will have taken over entirely.

Curly is reported to have hauled out 1200 tons of gold and copper ore a month from Lucky Jim, sending it to the Tyee Company's smelter at Ladysmith, and racking up a fortune for the owners. After 1910, however, owing to financial difficulties, very little development work was carried on, and Lucky Jim dropped from the news until Quadra's great forest fire swept the island in the 1920's.

Inspired by the early success of the Lucky Jim, claims were filed in many parts of the island by individuals and by mining companies. Those that were worked to any extent lay in the Lucky Jim and Great Granite area and on land near Gowlland Harbour. None, however, were as rewarding during these two decades as the famous Lucky Jim.

Some excellent farms were developed on the island.

Thomas Noble, the island's representative for the British Columbia Farmers' Institute in 1912, came to Quadra from Lancashire, England, in 1910, bought Dick Hall's property at Quathiaski and worked it into a well-equipped, up-to-date farm including livestock of cows, pigs and poultry. His twin sons logged on the island and later bought up large areas of interior and waterfront property for real estate development.

George Pidcock's diary during the early part of the century tells of numerous successful hunting expeditions with his brothers and with Billy Assu. Pidcock's sister Mary was also a crack shot. Danger from wild animals was no longer a problem; there was an occasional cougar, but most of the 800 cougars slain by Mary Pidcock's husband, "Cougar" Smith, were taken from the surrounding islands.

The Pidcock brothers opened their cannery at Quathiaski Cove in 1904, but sold out in 1906 to Nova Scotian T.E. Atkins who enlarged it and supplied it with new machinery. Atkins was also a partner in the drugstore chain of McDowell-Atkins-Watson Company and connected with various timber and sawmill industries and real estate ventures. The new owner brought in salmon from outlying stations in his gasoline launch, and engaged a Mr. Blackall to manage his store. Two years later, Quathiaski Packing Company bought out Atkins, changing the name to Quathiaski Cannery.

The Quathiaski Cannery burned down on August 31, 1909, in a dramatic fire described by John Antle: "We were tied up at Quathiaski Cannery wharf. About two o'clock in the morning the cannery was discovered to be in a blaze and we woke up to find the flames roaring over us and the paint frying on our boat and deck gasoline tank. Dr. Kemp and I cut and shipped lines as soon as possible but the vacuum under the cannery created by the fire was so great that our united efforts could not move the boat from the wharf. But Engineer Evans was busy with his engine and in the nick of time the welcome puff puff was heard, and I was glad to escape to the wheel house and give the bell that set her forging ahead. I do not think the boat could have remained at the wharf another minute without catching fire and in all probability exploding the gasoline tank. But a merciful Providence watched over us, and we escaped, the boat with some blistered paint and ourselves with a few burns and bruises."[39]

The cannery was rebuilt the next year and in 1912 was transferred to Quathiaski Canning Company Ltd., whose main shareholder was W.E. Anderson, with W.H. Malkin of Vancouver also holding shares. Anderson was the operator and the firm was known locally as Anderson's cannery. Anderson aided the research that uncovered the life cycle of the salmon when he reported that sockeye he had traced from the Salmon River running past his cannery, were heading for the Fraser, the only stream below Cape Mudge frequented by sockeye. He believed they came from Johnstone Strait, as they differed from Phillips Arm sockeye. Dr. Gilbert of the Fisheries Department was able to confirm this, and further studies by him based on this information resulted in the present knowledge of the salmon life cycle.

Anderson's cannery, like Draney's cannery and others along the coast, issued metal tokens instead of cash as payment for fish. These tokens were issued in three sizes, varying in value from 1 to 100 fish and could only be cashed for goods at the cannery store, where prices were so high that fishermen could bring supplies more cheaply from Vancouver. It was the strong protest of Chief Billy Assu that persuaded Anderson to end the controversial system of payment through tokens. The high prices in the store were also responsible for the organization of the Quadra branch of the Farmers' Institute.

The impressive Cape Mudge Indian fishing fleet was

gradually building up through these years as Assu encouraged his village to "go modern". Urged on by missionary and agent, Assu had his own great house hauled down to the beach. Indians and whites give varying accounts of its disposal. Some say it was pushed out to sea like a Viking bier; others that it was dismantled and the logs either burned or shoved into the water. At any rate, it was sacrificed, and other band members followed suit in 1915, first building new modern homes on land behind the longhouses.

"My father called us four boys together," said Harry Assu, Chief Billy's son. "I remember he told us: 'It's not the potlatches that are going to count now, it's money; we're going all modern: new houses, roads, everything.' When we were older, he got each of us a gillnet boat; later we got seine boats."[40]

Many fishermen still used rowboats, usually with oars but sometimes with a sail. Easthope Brothers Ltd. of Vancouver had manufactured the first engine for fishboat installation in 1900, but for a good many years the engines were only used in lower coast waters. Powerboats were banned north of Cape Caution until 1924, so that Indians who could not afford them would be on an equal footing with the white fishermen in that area.

1900 saw the first big fishermen's strike, with both canners and fishermen organized. During the following 50 years, at least 30 different fishermen's organizations were formed, and there were over 40 strikes. British Columbia Packers Company was formed in 1902 and at once began to buy up canneries up and down the coast. Large companies made substantial profits in the period 1900 to 1909, known as British Columbia's golden age of salmon, but many small companies were unable to weather years when runs were poor, and were forced to close down or sell out to the big organizations.

During the years 1900 to 1920, the fishing industry

improved its machinery and packing methods, a start was made on improving conditions and prices for the fisher-man, and the large companies were seriously studying ways and means of conservation.

-8-

By 1920, the white population of Quadra, massed mainly in Heriot Bay and Quathiaski Cove, had grown to almost its present size, while the Indian population at Cape Mudge was much larger than it is today. New schools were built to accommodate an increasing popu-lation; around 1921 a new and larger Valdes Island school was built on the original site, while in 1924 a school was built at Gowlland Harbour. The new com-munity centre (Valdes Island Social Hall), built in 1919 on a site midway between Heriot Bay and Quathiaski, served islanders from both areas.

An unresolved murder mystery that took place in the early 1920's is still talked of on Quadra Island today. Captain Dunn of Bold Point called at the Surge Narrows store, bought a supply of groceries and set off for home in his boat. The boat was found drifting off the point, with Dunn's body dragging behind it, one leg tangled in the boat's painter. There was blood in the boat, and the groceries and the boat engine were missing. Dunn, a peppery, outspoken individual, was said to have many enemies, but the mystery of his death has never been solved.

The advent of the gas engine had seen a reduction in fishing by dugout canoe and rowboat, but it was after motorboats were allowed in the Nass and Skeena Rivers and Rivers Inlet that the fleets of canoes travelling under sail began to disappear completely. Soon, fishermen would no longer haul in wet nets by hand from open boats exposed to all weather, as sturdy fishboats equipped with winches and comfortable cabins were seen in ever larger numbers. A new era in fishing was evolving. A cannery right on the island, close to excellent fishing grounds, resulted in a build-up of Quadra's fishing industry. Fishermen had depended on packer boats to seek them out and carry their catch to canneries, and in the 1920's the packers were neither speedy nor properly refrigerated for long trips. With Anderson's cannery at the cove, many part-time fishermen fished only the Cape Mudge waters and ran their fish in daily.

Mining and, to a much larger extent, logging continued as active industries until the fire of 1925. There have been three major fires recorded on Quadra: in 1896, 1919 and 1925, the last being by far the most devastating.

The 1925 fire started in Granite Bay, destroying the Lucky Jim mine-buildings and machinery. Fanned by a strong wind, it roared southward with terrific speed almost as far as Cape Mudge, demolishing the Gowlland Harbour school and numerous houses and buildings in its path. When it neared Main Lake, a prospector, H.N. Bacon, known on the islands as "Lord" Bacon, hurriedly travelled down as far as Gowlland Harbour warning residents that the fire was headed that way. The Dewey Vaughn family had moved to Gowlland Harbour only recently. Sceptical of the alarm, they hesitated until suddenly the smoke and flames bore down on them and they barely had time to collect their children and climb aboard the raft of Halverson, their neighbour and relative.

As they drifted down the harbour, watching the flames engulf their house, they took stock of the few belongings they had snatched up. Mrs. Vaughn had rescued only her best hat and her husband's new shoes. As the overloaded raft rocked in the waves, the hat and one shoe slid into the water and were lost. The raft floated down to May Island and stuck there until a boat from the Duncan Bay camp came over and towed it to Gowlland Island. Here the two families took shelter in an old, abandoned farm-house where they lived until new homes could be built. The Moss family felt themselves safe in their home on Stag Island in Gowlland Harbour but the crown fire jumped across the water to Stag and they were forced to spend the night in their boat until the flames died down.

In Heriot Bay dwelt author Francis Dickie, famous for his vast library of books, his large stock of exotically labelled homemade wines, and his twelve cats. Dickie and his little French wife Suzanne saw the raging flames heading directly for their newly-built home, and strug-gled down the steep, rocky cliffside to their rowboat, carrying a few possessions hastily collected. In this in-stance, the literary couple found they had salvaged the cats, an Airedale dog, bookshelves, a dictionary stand and several washtubs filled with books. They were about to push off with these and abandon their home to the fire when, as if on cue, the Columbia Coast Mission boat *Rendezvous* hove into sight from around the point, with Alan Greene at the wheel. The *Rendezvous, en route* to a nearby settlement, had come upon the disabled Forestry boat and offered to tow her raft to the fire at Heriot Bay. The raft was equipped with pressure gas pumps and a hose which the Forestry men used to spray sea water on the ground, underbrush and trees around the Dickies' house, keeping the fire at bay.

Many of the tall, blackened skeletons of first-growth timber remained upright for years, gaunt spectres of the forest fire, showing up stark and unsightly above the new growth all through the 1930's and '40's. However, houses and buildings were gradually replaced. Classes were held in a settler's home until a new school was built in Gowl-land Harbour later that year. Some surface work was done on the Lucky Jim mine two years after the fire, but the golden days of mining were over, and logging had re-ceived a major setback, with much of the mature timber lost. Small logging outfits continued to work, but the fire and the prodigality of early logging had taken their toll.

The forest fire did not reach Cape Mudge, but in the early 1920's an equally burning issue disrupted the lives of the Wiwekae and of Indians all up and down the coast. The government decided to step up its rate of prosecution of Indians who broke the Potlatch Law, ending once and for all the troublesome practice that kept Indian children from school and adults from work, and was the cause of increasing demands on welfare funds. The great "Christmas Tree Potlatch" of Dan Cranmer of Alert Bay touched off the series of trials that ended potlatching forever for the Cape Mudge Indians and lost them all the potlatch regalia and valuable coppers which symbolized the wealth and prestige of the Lekwiltok.

It was just before Christmas, in December 1921, that Daniel Cranmer gave his potlatch, one of the largest, in terms of wealth given away, in Southern Kwakiutl history. It was held on Village Island, the home of relatives of Cranmer's wife, to escape the notice of the Indian Agent. Cranmer said: "People came from all over, from Lekwiltok to Smith's Inlet. The invitation was given to all the chiefs of all the tribes. . . . Three to four hundred men, women and children turned up."[41]

Hamatsa and other dances were given the first evening. The second evening Chief Assu of Cape Mudge gave Cranmer the rights to a dance with shells and the right to several names. In return, Cranmer gave Assu a gasboat and $50 cash. The list of Cranmer's gifts to guests is staggering. It included 24 canoes, pool tables, violins, guitars, four gasboats, dresses, shawls, bracelets, 300 oak trunks, sewing machines, button blankets, 400 Hudson's Bay blankets, 1000 basins, glasses, washtubs, teapots and cups, bedsteads and bureaus, 1000 sacks of flour and sugar. Money was also handed out to adults, and small change thrown to the children.

"Everyone admits that that was the biggest yet," Cranmer said later. "I am proud to say our people [Nimpkish] are ahead, although we are the third. So I am a big man in those days. Nothing now. In the old days that was my weapon and I could call down anyone. All the chiefs say now in a gathering, 'You cannot expect that we can ever get up to you. You are a great mountain.'"[42]

Treachery, however, attended the potlatch. An Indian constable was helping the dancers with their preparations, but secretly taking notes after each part of the performance, which lasted several days. Names and activities of all the participants were listed, and the notes turned over to the Indian Agent. Virtually all the Southern Kwakiutl were involved in one way or another. Sergeant Angerman of the Royal Canadian Mounted Police made arrests, and over 80 Indians were summoned to appear in court, some of them people of the highest standing in the Kwakiutl system, including Chief Billy Assu. Lawyers came from Vancouver to defend the Indians at the trial, which took place in Alert Bay in early March, 1922. The court consented to the lawyers' plea that chiefs who agreed to surrender potlatching paraphernalia and give up the practice should be released, and the Indians were given a month to make their decision.

Only people from three villages, Cape Mudge, Village Island and Alert Bay agreed to the proposition. The rest were sentenced at the end of March to prison terms in Vancouver.

The potlatch was already losing much of its old flavour, due to the need for secrecy. Kwakiutl chiefs got round the law by going from one village to another to pay debts and distribute gifts, instead of holding one large gathering which might alert the Indian Agent, but this meant eliminating the ceremonies which were the main reason for the potlatch. "There was no éclat attached to the one who gave it, and the affair would fall very flat," said Agent Halliday.[43]

Chief Billy Assu called a meeting of his people when the ultimatum was given. He told them it was useless to continue the fight to retain the old customs. Some protested; they had debts to pay, and a future without the pageantry and social significance of the dances looked bleak. To many of the older Indians who could not read or write, the idea of a substitution of legal papers for public announcements at the potlatch was meaningless.

But the threat of prison sentences was daunting, and few wanted their chief to suffer the indignity of imprisonment. "Assu had great respect from everyone. Prison would bring shame to him," explained nonagenarian Louise Hovell of Cape Mudge.[44] There was resentment at the relinquishing of these family heirlooms, but almost every family surrendered its treasures as ransom for the chief.

Through the efforts of Chief Jimmy Sewid of Alert Bay and various members of the Kwawkewlth nation, a portion of the regalia has been returned by the National Museum of Canada. Half a century after the confiscation, it is housed at Cape Mudge in a unique cedar museum funded by Ottawa. Eventually, books, films, tapes and photos will be added to make the museum a learning centre of Kwakiutl culture. Opening day was preceded by a huge potlatch given by Chiefs Sewid, Harry Assu and Jimmy Wilson of Kingcome to dedicate the museum's 29-foot totem carved in memory of Billy Assu. The traditional cedarbark ceremony marked the formal opening on July 29, 1979, with native dances performed before a Cape Mudge D'entsiq dance board bought back by the National Museum from the Museum of the American Indian in New York. Fifteen hundred pounds of barbecued salmon were supplied free by the Indians for the hundreds of guests.

There was still no regular ferry service for Quadra in the 1920's, but water taxis continued to help fill the gap. In the early 1920's, Harry Fernace ran a 35-foot boat with an opposed two-cylinder gas engine, something of a novelty in its day, and Jim McPherson ran the *Connie Mac* after he retired from fishing. Bill Hall skippered a 30-footer equipped with a five-h.p. Yale gas engine in 1925, running from the government dock at Campbell River to the cove. Frank Gagne and Dick Bryant (son of Black Jack) started a service around 1927 with three speedboats, the *Blue Heron*, the *Blue Goose* and the *Blue Swallow*. They also had two larger boats, the *Arendale* and the *Otlinto*. Henry Reimbeault ran a jitney boat, the *Osoyoos*, in 1929 and Jack Lewis and his two sons, who owned the little cove store down the road from the cannery store, operated a taxi service with two boats until the advent of the first regular scheduled service.

Excitement came to the island in 1927 when the S.S. *Northwestern*, northbound in a snowstorm about five A.M. missed the channel into Discovery Passage and foundered on the beach at Cape Mudge. Alex Mercier, former fire chief for Quadra, was a boy of nine then, living at the cape, and he ran down with the others to view the wreck. She was a huge ship and, with her lights shining through the darkness and the falling snow, he says she had the appearance of a city suddenly risen up from the sea. She had been *en route* to Alaska, laden with Christmas supplies of every sort: butter, flour, lard, cheese, hams, blankets, clothing, innumerable toys, leather goods, tools and a myriad of other luxury items. The animals trapped below deck were shot, and a power launch came to take off passengers and crew on the leeside. The ship itself was left untended for months.

Settlers and Indians swarmed about the *Northwestern*, coming from as far as Heriot Bay in rowboats and gasboats and by horse and sledge to load up. In the weeks that followed, storekeepers complained to police that the island residents had stocked their homes with *Northwestern* cargo and no one was patronizing the stores. The police made an inspection tour of the island homes and there was a great scurrying about to conceal loot in root houses and hay lofts, as no one was sure of the rules of salvage. In the confusion, hams were left hanging in the attic and ship's blankets waved in the breeze on the clothes-lines, but the inspectors departed without comment and without confiscating anything. Eventually, a salvage tug came and towed away the ravished *Northwestern.* Quadra Islanders enjoyed a bounteous Christmas that lingered long in their memories.

The start of the Depression years saw Quadra Island facing the end of its economic boom. As a result of the fire and of overly extensive logging, camps had closed down, logging rails had been torn up, the ties rotted away and the forests gradually took over. Small logging outfits continued to log the few remaining stands of mature timber, but the time was fast approaching when a complete halt would have to be called until the second growth matured.

The Lucky Jim mine, its buildings and machinery destroyed by the fire of 1925, rallied briefly the following year when R. Crowe-Swords of Vancouver supervised some work, but this soon ceased. In 1932, Crowe-Swords staked 16 claims in an area where radium deposits had been detected years before, but no work was done and the property was abandoned. It was much the same story with several other claims during the 1930's. Though today's mining maps mark most of the island as "small deposits possible", the heyday of mining on Quadra appeared to be over.

As for the fishing industry, it was said: "From a business viewpoint the year 1931 was such a bleak one in the Fishing Industry of British Columbia that there seemed nothing pleasing about it. Many canneries were idle and some closed permanently."[45] In 1936 the Southern Kwakiutl formed their own union, the Pacific Coast Native Fishermen's Association, and Chief Assu and his eldest son Dan went to Alert Bay for organizational meetings. The Fishermen and Cannery Workers' Union reorganized into the militant Salmon River Purse Seiners' Union, while the equally active Pacific Coast Fishermen's Union of gillnetters and trollers (which took in many Quadra Island men) was organized in 1937. The two established jointly a bi-weekly newspaper, *The Fisherman*.

The famous strike of 1938 was dramatized by a flotilla of seiners that sailed *en masse* from Quadra Island to Vancouver. It won the first contract for purse seiners to be formally signed by the Canners' Operating Committee. The fishboats assembled at Alert Bay, then sailed to the well-concealed harbour of Granite Bay on Quadra on September 23 to lay their plans. In military style, they appointed squadron leaders and labelled their squadrons one to four. The next day they moved to Quathiaski where the Indians met them and Chief Assu promised support. The fleet then sailed for Vancouver in battle formation, their flagship flying the Union Jack. Arriving at the entrance to Vancouver Harbour, they passed by the sightseers crowded on Prospect Point and tied up at the Canadian Fishing Company docks. In addition to the 60 boats at Vancouver, the historic strike had included 18 boats tied up at Quathiaski, nine at Sointula and nine at Alert Bay, a total of 96 boats—almost all the seiners in British Columbia.

British Columbia Packers, which had been systematically buying up the smaller canneries throughout the years, added Anderson's cannery at Quathiaski to its list in 1938, losing it in a fire three years later.

The 1930's were unlucky years for Quadra schools. A mysterious explosion in 1935 demolished the ill-fated school at Gowlland Harbour which had burned up in the 1925 fire and had been rebuilt the same year. In 1936 the Valdes Island School went up in flames, destroying all school records from 1895 to 1936. Granite Bay School was forced to close in 1937 and remained closed for ten years, unable to meet the required student quota. Two new school districts were formed in 1936: Quathiaski Cove and Cape Mudge. A building on Joe Bigold's property was to be used until the new cove school was ready, but this burned down the day before it was scheduled to open. W.E. Anderson, the cannery owner, donated land for the cove school that eventually was built. This school was the nucleus of the present Quadra School.

Not all the happenings in the 1930's were dismal. The attractive little church in the Cape Mudge Indian village was dedicated in 1932, with Chief Assu assisting in the dedication service. It displays a memorial plaque to the memory of missionary R.J. Walker. This is the church of the Indians, who helped to build it, making by hand the pulpit and pews. Since it is the only church on Quadra at present, a few residents from other parts of the island come occasionally to its services.

Though times were hard, living conditions were inexpensive on Quadra, and the island attracted many of the unemployed who threw up shacks along the southern coastline from April Point (known then as Poverty Point) past Quathiaski and around the cape to Dogfish Bay on the east coast. One large concentration was at Whiskey Point. Here, along with many of the settlers, they hand trolled laboriously for salmon in dugouts, skiffs and battered rowboats, referring to their oars as "misery sticks". When there had been a good catch, bottles of cheap whisky were brought over from Campbell River and the trollers gathered on the point to drown their worries, forget blistered palms and "shoot the bull". Whiskey Point was the polite term for what was popularly known as "B.S." Point.

Colourful characters abounded among the amateur fishermen of the 1930's. A familiar figure around Heriot

Bay and many of the islands farther north was "Six-shooter Brown", a former cowboy who always wore a big black Stetson cowboy hat and kept a pipe permanently drooping from one corner of his mouth. He ran a little boat with an old engine held together with chicken wire and cement, and, like many of the old-time fishermen and loggers, was fiercely independent.

Deer were still plentiful on the island, and most residents, risking a fine, canned venison to tide them over the winter. There were plenty of clams and oysters, and almost everyone raised vegetables and fruit. It was the lack of cash for clothing, school supplies, and rent or upkeep on house and boat that caused distress. Boat expenses had priority. Many houses stayed unfinished for years, with tarpaper tacked to outer walls, front steps missing, and boxes serving as kitchen cabinets.

There was no set pattern of government relief during the Depression. Monthly payments were given in some cases; sometimes men received orders for food or a small amount of cash; at other times, work on government roads was a prerequisite; single men were often required to go to relief camps. Cape Mudge fishermen were given five-dollar orders for groceries once a month.

Still, many Quadra Islanders look back on the 1930's with nostalgia. Necessarily, it was a simple life-style, dependent for entertainment mostly on dances in the hall and picnics on Rebecca Spit. Rusty Beech, son of Quadra pioneer Jack Beech of Gowlland Harbour (who took over the Haas property) remembers when people came to the hall from all the Discovery Islands and from Courtenay and Campbell River for Saturday night dances that never broke up until dawn, or later still if a southeaster blew up. At big neighbourhood gatherings on the spit the picnickers roasted potatoes, fish and corn and washed it down with gallons of homemade wine.

This was an unsophisticated era. The highlight of the

week could be one of the picnics that Frank and Irene Endersby held on the beach below their cliff-top house at Heriot Bay. As darkness descended, islanders congregated around a bonfire and listened to young Eiva Oswald singing "My Alice Blue Gown", or Irene Endersby's more powerful voice echoing across the quiet water with "The Indian Love Call".

Irene was a former nurse and, as there was still no resident doctor on the island, she was sometimes summoned for emergencies. One that proved unnerving was a call to Quathiaski Cove to treat a man who had "cut himself". She arrived to find her patient lying dead in a pool of blood in his bathtub; he had cut his throat.

Heriot Bay's post office was still the living room of the old house that Hosea Bull had built by the esplanade. The Union Steamship vessel would blast out an alert at two A.M. as she circled to come in to the wharf. This was the signal for elderly postmaster Edgar Crompton to roll out of bed, haul on trousers and sweater and crunch his way in the dark across the wide crescent of pebbly beach to the wharf, two heavy mail sacks slung over his shoulders. Crompton, who had spent 16 years in Alaska, trapping, trading and goldmining, was postmaster in Heriot Bay for over a decade, until the post office was moved to Calwell's store on the waterfront, in 1945. This store, a landmark for nearly 40 years and owned latterly by the Dowler family, burned down in 1978 despite the attempts of three fire-engines to quench the flames with water pumped from the sea.

Calwell's store back in 1932 was an odd building on stilts, clinging to the side of the cliff and reached by climbing a long, steep stairway. Later, this building became the home of the United Church minister, Manly F. Eby, his invalid wife and her sister, Ethel Stephens. Eby, a frail little man, possessed an indomitable spirit and ventured forth in all weathers in his small boat, which eventually was driven up on the spit in a southeaster and wrecked. Ethel, a tall, sturdy woman, unfailingly cheerful, managed the large garden, sawed logs on the beach with a crosscut saw to keep them in firewood, and acted as cook, housekeeper and nurse to her bed-ridden sister. Now over 90, Ethel lives alone in a newer home, still tending her house and garden unaided.

In the late 1930's the steamer's new schedule brought her to the bay around noon, and at the sound of her whistle Heriot Bay residents streamed from their homes and converged on the wharf to exchange news and gossip. The Hundleys owned the Heriot Bay hotel at this time, and their two handsome daughters were an arresting sight among the crowd on the wharf. Only 18 and 15, dark Gladys and fair Beatrice were tall, generously-endowed young Amazons who, their father boasted, had done all the ploughing on the farm that had been their previous home.

Community dances at the hall involved whole families, from grandparents to small children and both the white and Indian populations. Medleys (or "Paul Jones" dances), known as "Brownies", were always included among the dances and successfully integrated old and young, Indian and white. Through the years there has also been a fair amount of intermarriage between the races. The old movies in the first, small Cape Mudge community centre attracted both whites and Indians. Settlers slid down the steep trail to the flats and crowded onto wooden benches in the hall. However, a form of segregation was self-imposed by both groups, who chose to occupy separate sections of the room.

Chief Billy Assu of Cape Mudge received the first of two royal honours given to him during his lifetime when he was awarded the Coronation Medal by King George VI in 1937 for his work among his people.

In the 1930's there were no car ferries to bring in the

tourists, no large shopping centres in Campbell River to entice residents off the island, and next-to-no cars, apart from the taxi of Joe Calwell. (Roads on the island were still unpaved and the rocks and ruts too damaging to engines and chassis.) Families like the widowed Mrs. Charles Gow and her four small children walked long distances across the island to pay calls, pausing often to nibble blackberries from the roadside bushes. Boats travelled from one island to another equipped as dress shops, beauty shops or radio repair shops, and islanders came down to the wharves seeking their services as pleasant and unexpected breaks in their daily routines. During these years of comparative isolation, Quadra Islanders grew to know one another, to depend on each other in times of stress, and to share their hours of recreation.

-9-

For half the years of the 1940's, war raged in Europe, drawing away many of Quadra's young men, some never to return. At home, the islanders battled with other problems, facing in the decade a series of catastrophes and ending with several progressive achievements.

A number of events had an impact on the fishing industry. Although the war created a demand for canned salmon, a shortage of employees and the dangers of wartime shipping limited the response. In 1942 Ottawa decided to evacuate all Japanese from the Coastal Defence Area, and Japanese fishing vessels, including 68 seiners, 120 trollers, 860 gillnetters, 148 packers and 141 codfishers, were taken into custody by the federal authorities. The British Columbia Packers cannery at Quathiaski burned in 1941, at the height of the salmon run. Fire broke out in the lighting plant, destroying the $60,000 building and causing another $60,000 loss when 400 cases of salmon and many cases of empty tins also

went up in smoke. The cannery, which had been acquired by British Columbia Packers only three years previously, was not rebuilt, a great loss to the island fishermen and the women who worked in the cannery. A dock
and net house, however, were provided by British
Columbia Packers for use of the fishermen.

Peace was declared in 1945, but the next year the island was rocked by an earthquake, considered as severe
as any experienced in Canada. Dr. Ernest R. Hodgson,
seismology chief at the Dominion Observatory, Ottawa,
came out to tour the islands, including Quadra, to estimate the damage and pinpoint the quake's epicentre. He
decided it was an elongated one reaching from Deep Bay
to Campbell River, mostly in the water near the shore.
Fisherman Dewey Vaughn and his wife were on the float
at Heriot Bay when the quake came and could see Rebecca Spit from where they stood. As the big wave struck
it, trees fell, and when the wave retreated they noticed
that a large part of the tip had disappeared. Three acres
of the point had sunk into the sea. Two long fissures had
opened up, dropping the land several feet below each.
The fir trees that grew thickly in the vicinity of the
fissures all died subsequently, and the dead trees still
stand, a grim reminder of the quake.

The congregation of St. John's, the only church on the
island built for the settlers, found it impossible to supply a
resident minister during the war years. Until Quadra
School was enlarged the church was used as a school for
primary grades, but thereafter it stood idle and forgotten.
Twenty years later the property was sold and the church
was torn down.

The cove school was renamed Quadra School when the
island came under Campbell River School District 72, in
1946. High school students were to attend Campbell
River High School, transported across Discovery Passage
on the island's first scheduled ferry, the *Victory II*.

Captain Bill Hall, who had been a Gowlland Harbour resident, formed the Campbell River Navigation Company which operated this 30-passenger, government-subsidized ferry. Dynamic Captain Hall had received his towboat master's ticket at the age of 21. His *Victory II* was a small ferry for passengers only (Quadra parents complained that some of the high school students had to stand outside on the boat deck in bad weather, on the morning and afternoon student crossings) but it was a welcome convenience for the islanders who used it during non-student trips.

Quadra School was one of the first to integrate Indian and white children. In January 1949 eight Cape Mudge Indian students from the seventh and eighth grades enrolled at Quadra School, and by 1958, as new rooms were added to house them, all the school children from Cape Mudge village had been transferred. The new school that had been built on the reserve was put to use as a kindergarten for the Yaculta children.

The Cape Mudge fishing fleet had grown by now to 40 gillnet and five seine boats. Assu and his family were also working for other advantages for their people. A complaint of Indians working off the reserve,—that they paid taxes but were denied the vote,—led to an amalgamation of two Indian unions, when the Pacific Coast Native Fishermen's Union joined the Native Brotherhood of British Columbia in 1936 to gain added strength for protest. In 1949, British Columbia granted Indians the vote and Assu's son Frank and William Scow of Alert Bay went together to Victoria to thank the Legislative Assembly. Assu's son Dan travelled up and down the coast studying Indian problems and made several appearances before federal government officials to state grievances. The deaths in the 1950's of these two sons of Billy Assu were a great loss to the Wiwekae. Harry Assu is the only surviving son of the chief.

After a series of amalgamations during the 1940's, the unions of white fishermen were also united in 1945 into one: the United Fishermen's and Allied Worker's Union.

Crown Zellerbach entered the logging picture on Quadra in 1949. The government signed over 60,000 acres, roughly two-thirds of Quadra Island, to the logging company as Tree Farm Licence Number 2. Under the terms of the agreement, the government retained the right to dispose of any timber within the tree farm until 1984, after which the big logging company will have complete logging rights. Until then, it is restricted to cutting stands of diseased timber, and thinning. Small companies are busily clearing out what remains of the marketable timber during the restriction period.

The 1950's saw the first signs of another radical change in the life-style of the island. A second passenger ferry simplified travel across Discovery Passage to Campbell River. There, the newly constructed John Hart Dam and the Duncan Bay pulp mill just outside Campbell River were the start of industries that altered that little village into a rapidly developing town with economic and entertainment attractions for Quadra Islanders. The opening of Quadra's Rebecca Spit as a public park, on the other hand, began to draw an increasing number of tourists in the other direction, to Quadra Island.

Construction of the dam began in 1947; the Duncan Bay mill started up in 1956. With industry down almost to zero on Quadra, employment in Campbell River beckoned. Some residents moved to homes across the passage, while others commuted on the little ferry, *Victory II*. Bill Hall had sold her along with his company in 1957 to Gerald Thomson of Heriot Bay who renamed the company Thomson Tug and Barge Ltd. Thomson announced that a barge would be available to ferry cars to and from Quadra Island.

Few took advantage of this last offer, for ferrying one's car across by barge was an unforgettable experience. First came the task of locating Thomson, who was apt to be anywhere along the coast with his tug. Then came the appointment of a day, the hour depending on the tide and apt to be at four or five o'clock in the morning. If a car owner preferred a more reasonable hour, he might arrive at Quathiaski to find the floor of the wharf lying a considerable distance above the deck of the barge. Undaunted, Thomson would prop two planks at a steep angle between barge and wharf and the passenger was invited to drive his car on. If his nerves were steady he might manage this, easing his car along, something in the manner of a tightrope walker on a guy wire.

The *Victory II* was replaced in 1959 by the government-subsidized ferry *Uchuk I*, capable of handling more than three times the number of passengers as well as seven tons of freight. She worked the passage on a three-shift basis, shift captains being Gerry Thomson, Tom Hall of Quathiaski Cove and Rowley Grafton of Heriot Bay. The *Uchuk I* was also detailed to take the high school students to Campbell River, while the little *Victory II* was scheduled for special trips to and from Quadra on movie and shopping nights.

1959 also saw the last four vessels of the Union Steamship Company turned over to Northland Navigation. The days of this pioneer company that figured so largely in the early history of British Columbia were now at an end.

Rebecca Spit, a three-quarter mile long strip of land that curves westward to form sheltered Drew Harbour, had been a Wiwekae reserve, but the land was exchanged by the government for the Wiwekae's present Drew Harbour reserve when the area was needed for World War One gunnery exercises. Later, it became the property of the Clandening family. On a walk to the end of the spit, always permitted by the obliging Clandenings, not a living creature was encountered in those days except for protesting crows and squirrels in the branches of the huge fir trees. In 1959 the spit was sold back to the government by the Clandenings and on June 20, 1959, was officially opened as a Class A public park, its natural, scenic and historical features to be preserved, with no commercial or industrial exploitation permitted. A plaque, orginally inserted in the trunk of an old fir tree, but now on a separate stand, states:

> This area was the property of Mr. and Mrs. J.C. Clandenning [sic], the site of many picnics in pioneer days. Through the years the people of Quadra and the surrounding islands have met in community gatherings, and as the Clandenning family graciously threw open their grounds for use on these occasions it is fitting that this area should now be dedicated as a public park for the use of the people at all times.

Rebecca Spit was named by Captain Pender of the *Beaver*, around 1864, after the British trading schooner *Rebecca*, which was engaged on the British Columbia coast for a number of years. Until the advent of the car ferries, the spit was still an ideal spot for picnics, or walks under the tall Douglas firs, along sequestered trails through undergrowth of fern, salal, Oregon grape and Indian pipe, or in open areas among daisies, summer bluebells and buttercups. The flat grass meadow halfway up the spit was used every summer for holiday sports as in the past. Rebecca Spit Park has been called "probably the most beautiful on the Gulf Islands" with its outlook on wooded islands and the mainland mountains. No one foresaw the effect the car ferries would have on the spit, or that steps would have to be taken to combat the problem of overcrowding.

One welcome result of the building of the John Hart Dam was Premier Byron Johnson's announcement in 1951 that electricity would be supplied to 174 customers on the south end of Quadra, at an estimated cost of $114,850. It would involve the laying of more than a mile of submarine cable across Discovery Passage from a point north of Campbell River spit to south of the April Point Yacht Club. The energy would come from the dam in Campbell River and would service Quathiaski Cove, Heriot Bay and its hotel, the Cape Mudge Indian settlement, and buildings at April Point. Before this, Cape Mudge and the Heriot Bay hotel had run their own power plants. Certain other homes and establishments had been hooked up to the hotel plant, but the power it generated was weak and erratic.

There were problems involved, however, after the cable was laid the following year by British Columbia Packers' boats. The cable was too short. Where it emerged, it had to be raised several hundred feet out of the water up a steep cliff. There was no spare slack for this, and it had to be pulled tight, emerging some distance out from shore, where the tidal movements swung the cable back and forth, eventually causing leakage and shorts in the circuit. It was decided to replace the submarine cable with overhead lines farther up the channel.

Near Menzies Bay, four power cables were strung across Seymour Narrows, reaching Quadra via Maud Island. Bright balloons were attached to the wires as a warning to aircraft, but it was a ship that blundered into them. In 1962, the American converted freighter *Cuff II*, *en route* to Alaska, was carrying an oil-drilling rig from Galveston, Texas, with a boom 150 feet high. The hydro cable was originally strung 165 feet above low water, but it had sagged somewhat in the centre, and unfortunately, *Cuff II* headed for the centre of Seymour Narrows. The great boom snagged the four power lines, ripped them apart and pulled down the wooden supporting towers. Afraid to halt in the strong tide, the ship continued on through the narrows, festooned with bright balloons and trailing long cables behind her.

Whenever the telephone cable was raised for repairs, it too was found to be well tangled with fishing tackle that had snagged on it over the years. In 1954 the cable was found to be unserviceable, and while a new one was being manufactured, service was carried on by radio-telephone. The North-West Telephone Company placed a small switchboard on the premises of B. C. Packers and provided three operators to man it; later, Frank Endersby operated it from his store on the Quathiaski wharf.

In 1955 a start was made on a project long hoped for, not only by Quadra Islanders but by all who sailed the Inside Passage route through treacherous Seymour Narrows. From the days of Captain Vancouver, sailors had spoken with awe of the whirlpools and boilers caused by Ripple Rock, hidden from sight in the centre of Discovery Passage. When added to the difficulties caused by the funnelling of waters into the narrow channel at flood tide, Ripple Rock made the spot a menace dreaded by all mariners.

On June 15, 1875, the U.S.S. *Saranac* from San Francisco, a paddle steamer headed for the seal islands, had foundered on Ripple Rock. Following this accident, over 100 torn hulls and 114 drowned men were reported victims of the rock. In 1884, H.M.S. *Satellite*, a 1420-ton corvette, steaming at 13 knots, was swept into the centre of the channel. Great upright waves rushed at her; the ship hit the tip of the rock, tilted over, hung a moment, then freed herself with 40 feet of false keel lost. In 1927 the Canadian National's *Prince Rupert*, travelling in foggy weather, hit the rock, which drove into her rudder and locked it with the propellor. The S.S. *Cardena* managed, with amazing dexterity, to draw alongside and pull her to safety.

The Canadian Merchant Service petitioned Ottawa to

find a solution. Various methods, such as drilling from above, were tried, the only result being more tragedy when a workboat carrying men ashore from the barge anchored over the rock was sucked into a whirlpool, drowning nine men. At last it was decided to tunnel into the rock from below the bottom of the sea.

The start of the operation was in October of 1955. A 570-foot mine shaft was sunk from the bluff on small, rocky Maud Island overlooking Seymour Narrows. One hundred feet below the ocean bed, a tunnel was drilled, ending below the peaks of the rock. Two shorter shafts were sent up the twin peaks and then the ricks were honeycombed with "coyote" drifts, six-by-four-foot tunnels, into which were packed 2,700,000 pounds of high explosives. Hardrock miners diamond-drill-tested hundreds of holes to be sure of the shape of the rocks.

On Easter Saturday, April 5, 1958, at 9:31 A.M., Victor Dolmage pushed the button that touched off $500,000 of explosives loaded inside the two peaks. No one was certain what might happen. Fishermen feared a massive destruction of salmon in the passage. Residents for miles around were tense, fearing anything from broken windows to a tidal wave. When the explosion came, black rock, hissing gas and salt water rocketed into the sky for 1000 feet at the rate of two miles a minute. Tons of rock and water were thrown up, and about 47 feet of rock was lopped off the 375-foot peaks.

Fears were unfounded. When the *débris* had descended, the sea lay calm. The turbulent whitecaps and whirlpools were gone. Within a half-hour after the explosion, the Fisheries Department checked and found dead codfish but no evidence that salmon had been destroyed. The enterprise had cost $3,100,000, and there are still swift currents to contend with, but all who travel Seymour Narrows feel that the blasting of fearsome Ripple Rock was well worth the cost.

Fishing had now become Quadra's major industry.

With fishermen banded together under the aggressive
United Fishermen's and Allied Workers' Union, the
1950's saw a succession of strikes for higher prices. The
canners followed the example of the fishermen and
sought strength in 1951 by amalgamating as the Fisheries
Association of British Columbia. Over the years, fish
prices increased and companies prospered despite strikes,
particularly in seasons such as 1953 when the largest sockeye pack in 40 years was canned.

The Potlatch Law was removed from the Indian Act in
1951, too late for most Indians, who had forgotten their
ancient songs and ceremonies. Chief Billy Assu, who had
given up potlatching 30 years earlier, was honoured
again with a medal in 1953, this time by Queen Elizabeth
II "for meritorious service."

The Indian fishing fleet continued to grow. A small
part of it may be seen on the Canadian five-dollar bill
which shows Harry Assu's seiner. The original picture
was taken by a provincial government photographer
during a big sockeye run in 1958 and shows the seiner
B.C.P. 45 making a set between Ripple Point and Bear
River. Harry Assu's son Mel was skippering the boat that
season. Shown in the picture are Mel Assu, skiff, Olly
Chickite, piling corks, Al Mearns (of Cortes) on the web,
Buster Seville, lead line, Andy Dick on the ring and Ron
Forrest (of Cortes) on the winch. The seiner in the background is the *Bruce Luck*, skippered by Don Assu.

-10-

The next two decades brought the tourists *en masse* to
the island, and both residents and government woke to
the need to alert themselves if they wished to preserve the
natural beauty, rural atmosphere and unique features of
the island. Conservation versus progress has been the
theme of this era.

The Cape Mudge village became the first Indian
village in Canada to seek municipal status. The fishing

skills of the Lekwiltok had made them a well-to-do reserve and, apart from Yaculta, they possessed the lighthouse property, leased to the government, and acreage at Village Bay, Open Bay, Drew Harbour and on Vancouver Island. Noted for their independence, they had built most of their homes themselves without government aid, and back in the 1920's had developed their own water and electrical systems. Since the mid-1950's their ruling body has been no longer one chief but a committee of three, elected democratically every two years. Assu, made an honorary chief, said with resignation: "There is no more glory for one man, nor is any one man higher than the next man."[46] Assu's son Harry was elected Chief Councillor on the first council, the other two being Sandy Billy and Mrs. Mary Lewis.

It was Oscar and Mary Lewis's son Lawrence, during the first of his four terms of office, in 1963, who first broached the idea of municipal government. Years were spent working on the plan. Municipal law was amended to permit Indians to vote on the issue, but a larger percentage of affirmative votes was required from Indians than was specified for white villages. After discussions involving the band council, reserve residents and provincial and federal representatives, the agreement was completed. Briefly, the draft promised the Indians full municipal status without the loss of their present rights. The council would consist of five members, four of whom would be Indians, and the mayor would also be Indian. It was put to the vote on January 17, 1972, and the result was just four votes short of the required percentage. Bitterly disappointed, Lewis resigned his leadership, despite pleas from Premier W.A.C. Bennett who wrote to him: "You were truly trying to break new ground for the Indian people.... We believe the leadership you showed has lighted some candles which will not go out."[47]

The law was amended two months later to bring the

plebiscite requirements into line with those of white villages, but so far no attempt has been made to hold a second referendum at Cape Mudge.

The opening of Rebecca Spit as a provincial park, and the introduction of a car ferry schedule between Quadra and Campbell River were the two events that touched off the tourist boom. On March 16, 1960, the new ferry, the all-steel *Quadra Queen*, was launched in the Allied Builders Yard in Vancouver. She could be loaded from either end, had a 14½ foot clearance for trailer trucks and could accommodate 15 automobiles and 99 passengers.

"Long life to the latest ferry, the *Quadra Queen!* May her twin screws pulse for many years at 10 knots to bring Quadra closer to the great world," said an editorial in *The Province* following the launching.[48]

Able now to take their cars across and travel in comfort in all weathers, more and more residents took jobs in Campbell River and commuted back and forth. Tourists seeking to explore new fields off the beaten track came in cars and trailers in ever-increasing numbers from Washington, Texas, Alaska, Alberta, Idaho, Oregon, California and a host of other locations, jamming the approaches to both wharves. On November 19, 1969, a new 30-car ferry, the *Quadra Queen II*, was put on the run and the first *Quadra Queen* started service as the *Cortes Queen*, running between Heriot Bay on Quadra and Whaletown on Cortes Island.

For a brief period the overcrowding was solved by the larger ferry, but with access now easily available to Cortes through Quadra, and news of the attractions of both islands spreading more widely among tourists through word of mouth and through Campbell River's advertising, long summer line-ups began to form again and Quadra commuters complained that the delays made them late for work.

Don Juan Francisco de la Bodega y Quadra, from a painting in the
Maritime Museum, Victoria.
(Courtesy of the Provincial Archives, Victoria, BC.)

"Skookum" Tom Leask, Quadra Island's Paul Bunyan, c.1890.
(Courtesy of Ed Joyce.)

Katie, Winnie and Roy Walker, Cape Mudge, 1897. Photo by
Hood. Hay was cut at Campbell River slough and brought to Cape
Mudge on a raft made with two Indian canoes, to feed the Walkers'
cow and calf.
(Courtesy of the Campbell River and District Museum.)

The Robert Yeatman home, 1899.
l to r: Robert Ernest Yeatman (on Maud); Samuel George Yeatman; Frederick Charles
Yeatman; Frederick Cook Yeatman; James Rayson Yeatman; Robert Wilson (neigh-
bour); Emma Rhoda Yeatman; Mrs. Robert Wilson (neighbour).
Photo by Hood. Yeatman collection. *(Courtesy of the Campbell River and District
Museum.)*

First Quadra Island Community Picnic, 1897. Welcoming the first member of parliament for Comox District. Drew Harbour, Valdes (Quadra) Island.
Back row, l to r: Alfred Joyce; unknown; the first Mrs. H. Bull; Walter Joyce; Aileen Hughes; Mary Hughes; Sarah Hughes; William Hughes; Robert Grant Sr., MLA.; Charlie De Verne.
Seated, l to r: Johnny Hughes; Harold Hood; Roy Hood; Hosea Bull with baby Cecil; Sam West; Mrs. W. Hughes; Weaver Jones.
(Courtesy of the Provincial Archives, Victoria, BC.)

Indian Agent R.H. Pidcock and his family, c. 1900.
Back row, l to r; Willie Pidcock; George Hugh Pidcock; Herbert Pidcock; Reg. Pidcock.
Centre row: Mrs. R.H. Pidcock; R.H. Pidcock; cousin or friend.
Front row: Mary Pidcock; Harry Pidcock; Margaret Alice Pidcock.
(Courtesy of the Campbell River and District Museum.)

Picnic at the Spit, c. 1902.
l to r: Herbert Pidcock; Mary Holmes; Aileen Hughes; Harry Pidcock; Mary Pidcock;
Black Jack Bryant; Mrs. Lillian Hood; Annie Davidson; Mrs. Davidson; John
Davidson; Willie Pidcock; Dick Hall; Roy Hood; Bill Bryant; Tom Yeatman; Rob
Yeatman; George Yeatman.
Right rear: Katie Walker; Mrs. Bryant; Mrs. Pidcock. Below them: Winnie Walker;
Alice Bryant. Below them; Daisy Bryant; Ray Yeatman; Mrs. Yeatman; Dot Yeatman;
Fred Yeatman Sr.; Fred Yeatman Jr.; Gordon Hood; Reg Pidcock; George Pidcock.
Photo by Hood. Yeatman collection.
(Courtesy of the Campbell River and District Museum.)

The Reverend John Antle and his family.
l to r: John Antle, Marian, Mrs. Antle, Victor.
front: Ernest. Victor was the child who went through the rapids.
(Courtesy of the Provincial Archives, Victoria, BC.)

Hamatsa emerging from the woods. From *The North American Indian* by E.S. Curtis.
(Courtesy of the Provincial Archives, Victoria, BC.)

Chief Billy Assu in regalia, c. 1910, in front of one of his house posts. Photo by Mrs. Sam Henderson.
(Courtesy of the Campbell River and District Museum.)

Hastings Outfit, logging. Granite Bay, c. 1908.
l to r: John Sandstrom; Emil Luoma; "Walleen"; Charlie Hodad;
unknown; Freddie Walker. Luoma collection.
(Courtesy of the Campbell River and District Museum.)

Lucky Jim mine, Granite Bay, 1911. This is "Curly", the No. 3 engine
of BCMT & T Co. (Hastings Outfit) at work at the mine. Photo by
Henry Twidle.
(Courtesy of the Campbell River and District Museum.)

The steamer *Cheakamus* at Heriot Bay wharf.
(Courtesy of the Campbell River and District Museum.)

The first Columbia Coast Mission ship, *Columbia*, launched in 1905
and in service 1905–10.
(Courtesy of the Provincial Archives, Victoria, BC.)

Constable Marshall's home, which was also the first jail, c. 1910.
There are still bars on the side windows.
l to r: Dorothy Yeatman; Mrs. Marshall. far r: Mary Smith, Cougar
Smith's mother.
(Courtesy of the Campbell River and District Museum.)

Heriot Bay Hotel, c. 1912.
l to r: Mrs. Bull; Hosea Bull; Mr. Jorgerson; Mrs. Scholfield; Rt. Rev. Bishop Scholfield
of Victoria; Mrs. Dick; 3 Jorgerson boys; hotel visitor; Jennie Millar and sister Mae;
Elva and Velma Anderson; Betty Law; Rev. Fred Comley; Mrs. Comley; Mrs.
Whipple (wife of storekeeper); Margaret Haslam (Heriot Bay teacher); Theodore
Spencer (Cove teacher).
(Courtesy of the Campbell River and District Museum.)

Anderson's cannery, c. 1912. l to r: cannery, net loft, China house.
(Courtesy of Alex Mercier.)

Anderson's cannery store, c 1912–14.
l to r: Jack Harper (manager); Elva Anderson; Mrs. Anderson; Mr.
W.E. Anderson; Velma Anderson.
(Courtesy of the Campbell River and District Museum.)

Valdes (Quadra) Island School teacher and pupils, c 1915.
Back row, l to r: Albert Joyce; Frank Yeatman; Helen Joyce; Jennie Rendle; Pearl
Wilson; Eddie Joyce; Vernon Summers; Miss Forester (teacher).
Front row, l to r: Alan Joyce; John Waters; Jane Wilson; Jack Rendle; Jennie Laclair;
Cecil Joyce; Blanche Rendle; William Waters; Clovis Wilson.
(Courtesy of the Quadra School Archives.)

Quadra Island Old Timers, August 1937.
l to r: Mrs. William Law, Sr.; Mr. R.J. Walker; Mrs. Alfred Joyce;
Miss Aileen Hughes; Mrs. Charles Gow.
Seated: Mrs. Bryant. *(Courtesy of Ed Joyce.)*

Francis Dickie moving back into his home after the 1925 forest fire on Quadra. *(Courtesy of Francis Dickie.)*

The big wheel of the Lucky Jim mine, all that was left after the 1925 fire. Photo by J.H. Wales, 1971.

Two house posts with cross beam, Cape Mudge village.
Photo by H.I Smith, 1929.
(Courtesy of the National Museums of Canada, Ottawa.)

Three grave posts, Cape Mudge village. Photo by H.I. Smith, 1929.
(Courtesy of the National Museums of Canada, Ottawa.)

Petroglyph, Cape Mudge village.
Photo by J.H. Wales, 1971.

Flotilla of salmon purse seiners on their way to Vancouver during the
1938 strike.
(Courtesy of the UF & AWU, The Fisherman.)

The *Victory II*, first passenger ferry between Quadra Island and Campbell River, 1949.
(Courtesy of Tom Hall.)

Ripple Rock explosion at 1:31 + 5 seconds, May 23, 1958.
(Courtesy of the Campbell River and District Museum.)

Cortes Island School, Manson's Landing, 1917.
Back row, l to r: Alice Marquette; Mrs. August Tiber.
Second row, l to r: Catherine Marquette; Mrs. Mary Marquette; Rose Manson; Anna Manson (Middleton).
Front row l to r: Ethel Tiber; George Marquette; Willena Smith (Thompson); Nicol Manson.
(Courtesy of the Campbell River and District Museum.)

John Manson and his nephew, Wilf, shearing sheep on Mitlenatch Island.
(Courtesy of the Ellingsens.)

Stag Bay, Hernando Island, with logging dump pier in the background, c. 1926. l to r: Ethel Hurren, Jane Manson, Mike Manson, Hazel Herrwig, Wilfred Manson behind Mike Herrwig, Ray Morris behind Ralph Morris, Nana Hurren.
(Courtesy of the Ellingsens.)

Mike and Wilfred Manson haying on Hernando Island, c. 1926. *(Courtesy of the Ellingsens.)* Hernando is no longer a logging or farming area but is owned by 40 shareholders who commute to their summer homes by plane or boat.

It was a vicious circle. To service the new crowds, the government announced it would spend $250,000 on the design of a new, $2 million, 50-car ferry for the Campbell River-Quadra Island run, using the 30-car ferry *Quadra Queen II* on the Comox-Powell River run. The seven-member Quadra Island Advisory Planning Commission sought professional advice, then suggested the *Quadra Queen II* would be adequate if sponsons (steel buoyancy chambers) were fitted to her hull, allowing more logging trucks to be carried as well as the full quota of cars. They gave as their reason for voting against the 50-car ferry that the new ferry would require expensive parking and docking improvements and that Quadra Island had a water shortage and would suffer from the increased demand if more people were encouraged by the larger ferry to come to the island. (In the previous year, the ferry had carried one million passengers, becoming the busiest Highways Department-operated ferry in British Columbia.) The offer of the 50-car ferry was withdrawn, and the *Quadra Queen II* continued her services to Quadra residents and to line-ups of cars and campers during the tourist season.

In 1965, the government divided the province into 28 Regional Districts, with Quadra Island lying in the Comox-Strathcona district. By 1970, regional and community planning had begun. The Regional Board decided in June of 1970 to impose land use controls on Quadra, and a lengthy survey of the island was published in 1971 by the Regional Planner of the district. It is the only study made by the planning committee on the region's islands to date. Its conclusion was that recreational development was the logical choice for Quadra. "In the increasingly complex and tension filled world most people are facing," it stated, "we cannot afford to lose places such as Quadra Island."[49] But the recreational angle alarmed many. Their contention was that an influx of tourists and summer vacationers to the island's seaside and wilderness lake areas was not the way to maintain their island as a refuge from the complex and tension-filled world.

Quadra worked out its own zoning bylaw which was approved by the Lieutenant-Governor-in-Council and adopted in May of 1975. The island voted against the proposal of the Regional District's planning committee to apply the building code to Quadra. Sixty residents chartered two buses and travelled to Courtenay to appear before the Regional Board and protest the motion. To quote Sam Hooley, a long-time representative of Quadra on the Regional Board: "When you try to put a saddle on an Islander, you've got trouble."[50]

The building code imposition was abandoned, and a siting bylaw, also opposed, was redrafted to include only Denman and Hornby Islands. Chairman Berry of the planning committee said that both Quadra and Cortes "appear to be active in policing their own building regulations. They're saying, 'we don't have any problems and we'd rather look after our own affairs.'"[51] Today, Quadra is attempting to formulate its own Settlement Plan to submit to the Regional Board.

The Quadra Island survey points out that fishing, logging and agriculture, once the main occupational pursuits of the settler, are no longer considered important to the majority, who are mainly workers who commute to jobs in Campbell River, a number of pensioners, and part-time summer residents. Pleading for recreational development, it indicates Quadra Island's attraction for tourists due to its proximity to Campbell River's urban services and markets. Quadra's 30-car ferry service gives easy access from Victoria and Vancouver, its sheltered coves provide safe anchorage for pleasure craft, and its long and varied coastline offers a broad range of recreational opportunities. Quadra has several resorts and marinas and also provides road and ferry connections to

Cortes Island, another popular area. Other assets are Quadra's parks and beaches and its 2400 acres of clear, unpolluted lakes, useful for domestic water supply, canoe travel and sports fishing. Quadra is located in the centre of a major bird migration route, the Pacific Flyway, and is an important stopover for a wide variety of waterfowl. Valuable salmon fishing grounds entirely surround the island.

The survey considers that with expanded facilities Quadra could be expected to equal Campbell River in the number of tourist camper nights. It suggests the success of Rebecca Spit Park might be repeated in other locations on the island, but after the 500-visitor camper nights recorded in 1968, with campers parked side by side in rows along the woodland paths, fears for the ecology of the island led to the closure of the spit to overnight camping. Visitors are now accommodated at the Wi-wekae Band's forested court, and lovely, wild Rebecca Spit is free again for picnics and secluded woodland walks.

The Cape Mudge minister, Ron Atkinson, wrote regularly for Quadra's paper, *Discovery Passage*, before his transfer elsewhere. In 1974 he wrote: "An islander is different from a mainlander. There is an island temperament. The true islander prefers being somewhat removed from the mainstream of society. He treasures detachment and insularity. He doesn't 'go over' more than he needs. The islander desires two things: solitude and simplicity."[52]

So Quadra Islanders continue their struggle to maintain their independence and to hold back the tides of What-is-to-be. When Percy Joyce, grandson of settler Alfred Joyce, was asked by *Discovery Passage* to give his reaction to plans for the island, he replied: "We're going to have progress whether we want it or not. . . . I'd like the island to be the way it was when I was a boy."[53]

CHAPTER 2

Cortes Island

Cortes Island is Coast Salish territory and long ago the Salish inhabited the island on both east and west shores. An old Indian told pioneer John Manson that the beaches of Cortes were once "black with Indians"[1] but the smallpox epidemic of 1862 killed off nearly all of them. The Indians, feverish from the plague, would stagger from their tents to the sea to cool themselves, and collapse and die as they crawled back up the beach. In later years, a different tribe of Salish came to occupy the island that smallpox and enemy raids had depopulated. Squirrel Cove is the principal village of the Klahuse Salish today, but they have come to settle there from Toba Inlet and Salmon Bay only since the arrival of the missionaries, in the 1890's.[2]

Raids by enemy Indian tribes were a constant menace in the past, particularly those of the aggressive Haida and Euclataw (Lekwiltok). John Manson's aged Indian informant said the long, rolling hillocks at Smelt Bay on Cortes were fortifications against attack, and he remembered raids there by the Euclataw. Other Indians have said that the high canyon walls of Gorge Harbour were used in the past by Salish tribes to roll large boulders down upon the canoes of raiders. Chief Billy Mitchell of the Squirrel Cove reserve remembered his wife's great-grandmother who died at the age of 114 and bore two long scars on her back from spears of the Haida. She was wounded when she fought, unsuccessfully, to prevent them from seizing her young son for a life of slavery.

Midway between Smelt Bay and Manson's Landing, just north of the old Indian village of Paukeanum a petroglyph can be seen on a huge granite boulder. Only visible at low tide, this shows the outline of a fish or whale, nine feet long. At Gorge Harbour, there are Indian designs half way up the perpendicular cliffs. To reach the spot, Indian artists must have been lowered by cedar ropes. The pictographs stand as mementos of these ancient dwellers on Cortes.

As Spanish and English explorers followed one another up the coast in 1792, inlets and islands often received both Spanish and English names but, except for Point Mary, named by Vancouver after his sister, Cortes and its adjacent small islands retain the names given them by the Spanish explorers, Galiano and Valdes. Cortes Island itself (more often spelled Cortez in early days) honours Hernando Cortes, the Spanish conqueror of Mexico, who attacked and overthrew Montezuma's Aztec kingdom of 200,000 with only 500 soldiers and 16 horses. The small island Hernando, to the south of Cortes, also honours this soldier. Marina Island, west of Cortes, was named after the beautiful slave, Marina, whom Cortes chose from among his captives at San Juan de Ulloa and made his mistress. Marina gave him useful information about the country, which aided the success of the invasion. Marina Island was renamed Mary Island from 1849 to 1906, after which the original name was restored. The Ulloa Islands (called Twin Islands on British Admiralty and American charts) are named after the scene of Cortes' first victory.

In July of 1792, Archibald Menzies, the botanist of Vancouver's party, "went on a small excursion with Mr. Broughton in his boat." This was to Squirrel Cove, where they landed and followed a stream of water to "a Lake in the Wood which was apparently filled at high water by the impetuous force with which the Tide rushes into these narrow Inlets."[3]

Captain Daniel Pender of the Royal Navy surveyed the British Columbia coast from 1857 until 1870, in a series of vessels: the *Plumper*, the *Hecate*, and the paddle steamer *Beaver*, hired for a time from the Hudson's Bay Company. While in charge of the *Beaver*, Pender named Von Donop Creek on Cortes after a midshipman on H.M.S. *Charybdish*, and Carrington Bay after an Admiralty draughtsman. Captain Richards also surveyed the coast during this period, and named Sutil Channel on the west coast of Cortes after Galiano's vessel. Reef Point, at the southern tip of Cortes, so called because of its dangerous, mile-long reef, had its name changed to Sutil in 1945. The Subtle Islands off Cortes have acquired an English translation of the Spanish name *Sutil*.

Whales abounded in Georgia Strait before the whalers came. As Vancouver left Desolation Sound and passed Cortes Island, he wrote in his journal: "Numberless whales enjoying the season, were playing about the ship in every direction."[4] The Indians of Cortes had their own way of whaling, very unlike that of the Nootka Indians, according to Chief Billy Mitchell of Squirrel Cove, who said his father told him the men would line up straight out from the shore at Smelt Bay. As the whales came close to them they moved out and encircled the pack, all the while dropping clam shells which frightened the whales in towards the shore. They kept dropping shells until the whales were driven up on the beach and lay helpless.[5]

Whaletown received its name from the whaling station that was operated on Cortes Island from 1869 to 1870 by the Dawson Whaling Company, owned by James Dawson. A news item in the Victoria *British Colonist* of February 24, 1869, stated that "Mr. Dawson's schooner *Kate* arrived from the Gulf of Georgia yesterday with 2400 gallons of whale oil, consigned to Lowe Bros. Mr. Dawson... designs establishing his station on Cortez Island... in the vicinity of which he anticipates a good catch during the coming season."

In June of the same year, the newspaper reported "Mr. Dawson from Cortez Island" as saying all was in readiness and that his boys would start out after whales in a few days. That month they secured five whales, averaging 80 barrels apiece, a total of 13,000 gallons of oil, in demand for soap, tallow or cooking fat, candles and lubricants, and "worth in New York City, this very day, $1.20 per gallon in greenbacks, or 87 cents in gold." The company killed three other whales but lost them; two were later picked up by other whalers.

In August the paper's headline read: "Splendid Results of the Operations of Dawson & Co.'s Whaling Party." The whalers had gone to Cortes to try out the blubber of two humpback whales, one of which was 51 feet in length. Later that month, Captain Douglass of the Dawson crew recorded three whales taken on a Wednesday, the whales having been killed in three-quarters of an hour from the time of fastening the line to them. On the Friday they killed one more, and lost yet another when it sank.

In those early days of whaling on the coast, small boats were lowered from the schooner when a whale was sighted, and the harpoon bombs were launched from the small pursuing craft. After a long and often dangerous struggle, the dead whale was towed to the schooner. Captain Abel Douglass, skipper of the *Kate*, described two weeks of whaling in August of 1869 in a letter the following month to the *British Colonist*. With three whales aboard, the *Kate* was forced by fog and a southeast gale onto the reef that extends out for several miles from Marina Island. There was a delay for repairs and then she set out once more. The two boats were lowered from the schooner, although a stiff southeaster was blowing again.

One boat struck a whale with a harpoon, and Abel Douglass in the second managed to put his iron in as well. The whale was "a rather hard fellow to kill." After three bombs were put into him, the great mammal circled and dived. He came up directly under the second boat, thrust at it, put his head on it and pushed it "a considerable distance" under water. The boat surfaced bottom up and the men "swam for their lives." The first boat secured Abel Douglass's line which had run out 180 fathoms, and pulled up to the whale, finding it dead. The catch was their first concern, and they towed the whale to the beach before picking up Abel Douglass and his crew and righting his boat. The whale was cut up, the blubber sent aboard the *Kate* to the tryworks at Cortes Island, and the whalers set out at once to look for more whales. Despite their harrowing adventure, Douglass wrote: "All hands are in good health and good spirits."[6]

The schooner *Kate* arrived in Victoria from Cortes on September 17 with 150 barrels of whale oil. To date, the station had produced 20,000 gallons of oil. There was no closed season for the whales; the whalers hunted the whole year round. It is not surprising that on December 23, 1871 the *British Colonist* reported: "Whales are getting scarce in the Gulf." Dawson and Company had removed their station from Cortes to Hornby Island the previous year, and by January 1872 the company was in liquidation. The schooner *Kate* and all the equipment were sold at auction on March 19, 1872.

No signs remain of the whaling days at Cortes. The old pier used by the whalers was across the bay from the present wharf and is long gone.

The explorers, the surveyors and the whalers came and went. The miners came, staked claims and departed. Japanese loggers came with horses and left behind traces of their cedar logging skids. Then came the settlers, turning to farming and hand logging, bearing the hardships of pioneer life on an isolated island, confident there was a future for them on Cortes. This is one of the most beautiful of the Gulf Islands with its placid lakes and rugged gorges, and its sandy beaches with an abundance of shellfish available for the taking.

The first settler was Michael Manson, after whom Manson's Landing is named. One of four brothers who came to Canada from the Shetland Islands, he arrived at Manson's Landing (known then as Clytosin) in 1886, pre-empting land there the following year and receiving his Crown Grant in 1913. With him came George Leask, with whom he had boarded when he had first arrived in Canada and had become for a time a merchant in Nanaimo. Leask and Michael's brother, John, both pre-empted land on Cortes in 1888. At this time there were only two other white men on the island, hand loggers working in Von Donop Creek. There were no roads or buildings, and no steamer service to or from civilization, there were no gas engines, and local travel was by rowboat, or by dugout canoe, driven by sail when winds were fair, propelled by paddle when they were not.

Mike Manson's romantic life includes the story of his marriage, an elopement with Jane Renwick, daughter of his partner in the Nanaimo store. Renwick was a tyrannical parent and Jane was desperately unhappy. Manson borrowed an Indian dugout canoe and carried her off in it to Victoria where they were married. They had 13 children. Half the family was wiped out in a diphtheria epidemic, but all of the "second family", which included two sets of twins, survived.

On Cortes, Manson built a trading post and bought a schooner to bring supplies up from Nanaimo. On the first trip up the gulf, the schooner was loaded with mixed cargo including a deck load of hay, six cows and a pair of oxen. Travelling north in a strong southeast wind, the vessel was passing Reef Point (now Sutil Point) when the centreboard, which had been down to prevent too much leeway, struck a boulder on the reef, though the hull stayed clear of the rock. The old charts had failed to show the reef extending out so far and they were lucky not to have wrecked the heavily-laden vessel in such rough seas. They got 24 hens and a rooster ashore and into a chicken house, but on the first night mink got into the house, sucked the blood of every chicken and left the birds lying dead under the roosts.

Mike's trade was mainly with the Indians, on the Squirrel Cove reserve on Cortes and also on the neighbouring islands. The Indians fished for dogfish with lines of cedar bark and traded the dogfish oil for supplies. Manson sold the oil to the coal mines in Nanaimo as a lubricating oil for 37½ᶜ a gallon in good times and 25ᶜ when times were bad. Bear gall, produced from the occasional bear that was shot on the island, was sold as a medicine to the Chinese working in the mines. Mike also traded for deer horns which, when soft, brought 25ᶜ, 50ᶜ and 75ᶜ a set, according to size. Mink skins brought 50ᶜ if good, coon skins 25ᶜ and marten about $1.00.

Manson extricated himself from a ticklish situation while delivering supplies to the Cape Mudge Indians. As he was leaving with his payment, $1748, one of them ran up, pointing to a dead Indian on the beach, and begged him to settle an argument that threatened to erupt into a gun battle. A number of Indians had gone down to Gastown (Vancouver's name at the time) and when returning with a keg of whisky and three cases of square gin, were forced to seek shelter from a storm at Sechelt. As they attempted to land in rough seas, their canoe overturned and the keg of whisky was lost. One group said the dead Indian had been drowned; others said he had been murdered by an Indian who blamed him for the mishap.

Manson examined the dead man and under his hair felt the skull knocked in by the handle end of a paddle. To reveal his discovery at this time might have meant his own death, for the Indians were armed with rifles and the accusers had sworn they would kill both the murderer and Manson as well if he obstructed them. They said both corpses would be thrown into the bay "because no one would know what happened."[7]

The accused man by now had hidden himself behind Manson for protection. Mike drew from his pocket an old insurance policy which he carried because of its impressively legal gold seal and ribbon and told the men sternly his paper said the Queen had appointed him as a powerful *tyee* (chief) who must be obeyed or the village would be burned and many would be hanged. This did the trick. The Indians capitulated and Manson collected the rifles and reissued them to ten young men whom he appointed special constables to guard the prisoner. He smashed the gin bottles and then spent an uneasy night in the chief's house, with the chief's wife guarding his money. In the morning, he told the Indians a coroner must decide the prisoner's innocence or guilt and sent him, along with witnesses and the dead man, by canoe to Comox. The accused man received a five-year sentence. As Justice of the Peace, Manson was often called to other islands to investigate breaches of the law, ranging from liquor offences to murder.

Manson's younger brother John did a phenomenal amount of rowing in those early days. When large orders came in from logging camps for fresh meat, he hunted

down cattle running wild in the bush, killed and butchered them on the spot, packed the meat to the nearest
beach and then rowed it to the camps. At other times he
would row his butchered beef and mutton down to
Comox. He kept sheep on tiny Mitlenatch Island, about
six miles from Cortes, transporting them at the start, two
at a time, in his skiff. Once he rowed 100 miles each way
to the head of Knight Inlet and back to bring out two
school girls to board at the Mansons' home and raise the
number of available pupils to the number required to
open a school.

Wolves were reported a pest on the island by Tom Bell
in his 1895 Agricultural Report. John Manson took
strychnine-poisoned chunks of deer meat into the bush
and impaled them on stakes driven into the ground to kill
off the wolves who preyed on deer and sheep. This was a
time-honoured method used by settlers and trappers in
British Columbia and Alaska. John lived on Cortes until
his death at the age of 88. Michael lived only a few eventful years on Cortes, leaving to become superintendent at
the Wellington Colliery Company at Comox, which was
owned by Lieutenant-Governor James Dunsmuir. He
was elected M.L.A. for the Comox riding in 1909 and
1912, then returned to the islands to farm on Hernando
and was M.L.A. for the Mackenzie riding from 1928 until
his death at 75, in 1932. The provincial government
bought the rocky little island of Mitlenatch that the
brothers had owned, and turned it into a nature park and
bird sanctuary.

There were 17 mineral claims recorded on Cortes near
Carrington Bay during 1898, where small quantities of
molybdenite in quartz veins in granite were found. After
this, no mining activity is mentioned in the British
Columbia Mining Reports.

Easier days came for the pioneers when the little Union
Steamship *Comox* was launched in 1891 and made her

first run of the up-coast logging camps the following year with a weekly stop at Cortes. The 1894 *British Columbia Directory* lists 40 names under Clytosin, about equally divided between loggers and fishermen. Settlers farming at Whaletown in 1898 according to the Voters' List were William Burridge and Henry Corby, with Willis Nendick ranching there and Laurence Rose a Whaletown general merchant. William Viscent farmed at Smelt Bay; at Clytosin there were farmers Andrew Halcrow and Grosvenor Miles, engineer Horace Heay and his brother Alex, traders Michael Manson and his brother John. Frank Vaughan had a ranch at Gorge Harbour, and listed as "farmer, Cortez" are Sam Coulter, Andrew McNeel, William Seaton, Peter Shufer and Sam Thompson, with James Trahey a rancher. Early residents of the 1890's whose relatives continued to live on Cortes for many years included August Tiber and R.R. Allen.

The population had increased to the point where a school was needed, and in August 1895 a log building near Gunflint Lake, owned by Alex Heay, was opened as the island's first school. There were 12 pupils enrolled and Mary E. Ward was engaged as teacher with a salary of $50 a month. Later, school was held in a small log building at Clytosin, and still later in a log building owned by Manson.

In the Agricultural Report of 1903, Nicholas Thompson of Whaletown wrote: "We have got one more settler on the island this year and a school started. The one drawback to us at present is we have no wharf. . . . There is plenty of land for pre-empting around here, but it is hard clearing. We have a boat call twice a week with mail, and there is a good market for all we can raise. There are twelve ranchers on the island, men, women and children, 60 all told. Between 150 to 200 acres of cultivated land."[8]

Pioneers on all the islands put time and energy into acquiring schools for their children and keeping them open. John Antle reported in 1908: "Cortez Island is at last fortunate enough to have a school teacher. She resides at Mr. J. Manson's."[9] The Public Schools Report lists a school at Whaletown in 1910. Miss H.M. Bland, the teacher, received $60 a month and the trustees were J.P. Allen, secretary, N. Thompson and Mrs. D. Robertson.

Cortes was called a lumber camp in the *British Columbia Directory* of 1898, with logging companies at work, and local farmers supplementing their incomes by hand logging. The Fraser River Saw Mills company moved to Cortes in 1908; logs were loaded by ocean freighters that called at Coulter Bay. The supply of timber seemed never-ending to the pioneers. Ned Breeze, a Cortes pioneer who died there in 1948 at the age of 90, cleared land on his Gorge Harbour farm by boring a hole in the trees and setting a fire in the hole to burn them down. He maintained he never used an axe or saw; no one felt the need to utilize all the timber.

Evan Humphreys, a former resident of Cortes, says his grandparents, the Percivals, came from Liverpool to make their fortunes in Canada and ended up on a chicken farm at Gorge Harbour on Cortes. Their orchard still remains, but the site of their home is now a marina. Evan's parents told him of early days on the island when nearly everyone operated a still and there was a great scramble to conceal the evidence when the police boat was sighted. There were a good many Scandinavians logging on the island and they patronized the dances at the Gorge Harbour hall. When sufficient home brew had been consumed, one of the loggers inevitably leapt on a bench and shouted the time-honoured battle-cry: "Ten thousand Swedes ran through the weeds, chased by one Norwegian!" and the warring factions exited in a mob to establish supremacy in a glorious free-for-all.

The telephone line crossed Cortes in 1910 when a line

was laid for Quadra Island from Sara Point on the mainland. On Cortes, only stores were hooked up to the line at first, at Whaletown, Manson's Landing, Squirrel Cove and Seaford.

In 1907 the little Union Steamship *Comox* ran aground on Cortes Island reef in a heavy fog; passengers were carried ashore in small boats and then taken to Heriot Bay on Quadra and transferred to the *Coquitlam*. A three-foot hole between the boiler and starboard bunker had to be patched on the *Comox* and she was towed back to Vancouver by the tug *Tartar*. Later, the luxurious "loggers' palace", the *Cassiar*, took over the run, replaced in turn by the twin-screw steamer *Cowichan*, which was easier to manoeuvre in narrow channels.

During the years preceding World War One, Alan Greene of the Columbia Coast Mission was living in a cabin overlooking the harbour at Whaletown. He visited his congregations by travelling in his little 14-foot open gasboat and, he says, "apart from the occasional sinking of my craft, had a very happy time working from Whaletown up through the Yaculta Rapids as far north as Sayward."[10] When the snow came, he took to the steamers, the *Queen City* and the *Cassiar*.

A government surveyor reporting to the Minister of Lands in 1916 called Cortes Island well settled and better served with roads and trails than any of the neighbouring islands. He found the chief settlements to be around Manson's Landing and Whaletown, Manson's Landing having a Farmers' Institute co-operative store, a post office, telegraph office and school, while at Whaletown he noticed a general store and post office. He mentioned that sheep were being raised successfully by John Manson.

After the rush to settle on Cortes at the start of the century, population figures began to decline. The loggers were a shifting population and seldom stayed to settle.

Cortes was given publicity when the Union Steamship Company ran a series of weekend cruises to Savary Island, Toba Inlet and Cortes, a 360-mile round trip costing $12, including meals. During the 1930's, however, there were many abandoned homes and fruit farms on the island. In the postwar era, settlers began to come to Cortes again. Logging had removed the cream of the timber crop; some saw tourism as a replacement industry, but others hoped Cortes would retain its secluded charm.

Unlike Quadra, Cortes is an island of churches. In the 1950's there were three: St. John the Baptist at Whaletown, St. James at Manson's Landing and St. Saviour's-by-the-Sea at Cortes Bay. Before it burned in 1948, there was also a small church at Squirrel Cove. At present there is a little Catholic church in the Indian village there. The newest church, St. John the Baptist, built in the 1950's by residents and the Columbia Coast Mission, replaced the first small frame building that had also served as a community hall. At this time it was decided to locate school, post office and store in the vicinity of the church, and all were located in Whaletown as a central area most convenient for settlers. The mission provided a weekly medical and dental clinic. There were no crowded subdivisions, the island had maintained its wilderness appeal, and it still offered low-cost living. All that was needed was local transportation.

This came near the close of 1969 when Quadra Island acquired a larger ferry and the 16-car *Quadra Queen* was rechristened *Cortes Queen* and placed on a run between Heriot Bay and Whaletown.

In 1970, electric power was brought to Cortes to serve the 150 residents who had been dependent over the years on kerosene and hurricane lamps. A 32-mile distribution system through submarine cables supplied the power from Sara Point on the mainland to Mary Point on

Cortes. The lowest point in the underwater crossing was more than 1500 feet, the deepest of any submarine circuit in British Columbia's hydro system to that date.

Trouble arose in 1973 when wolves were seen on Cortes, thought to have swum over from the mainland. Rancher Ken Hansen reported 13 sheep and his best ram lost to wolves in one year and said he was selling off his remaining sheep. Others said deer bones littered the island wolf trails, and horses were "spooked" by the smell of wolves on the horse trails, which the wolves had taken over as their own. Conservationists declared the wolves were an endangered species and should be spared, while ranchers and farmers felt the wolves should be eliminated and the sheep and deer protected.

In 1974 the provincial government set aside 130 acres at Carrington Bay for a recreation area, and 117 acres at Manson's Landing for a Class A provincial park. The park includes 4,000 feet of beach facing the lagoon and 1300 feet of white sandy beach at Hague Lake. At Smelt Bay there is a 40-acre Class A park. This park, its name taken from the fish that swim from the sea into the creeks to spawn, offers a fine example of the smooth, sandy beaches of Cortes. Here also are seen the rolling, green hillocks that once were Indian fortifications.

In Gorge Harbour a marina and campsites welcome the boating tourist, while Squirrel Cove has a small store and post office. Its sheltered waters attract a crowd of summer holiday boaters to its government float. Bringing frequent visitors to the island is the Cortes branch of the Cold Mountain Institute, which has 23 acres on the southern peninsula near Smelt Bay. Industries on Cortes today consist of shake mills, an oyster co-operative and considerable craft work.

Population on the 15-mile-long island has grown to over 500 and appears to be steadily rising, though the Regional District Board has announced as its policy an opposition to bridges linking it to other islands, or further ferry service that might increase demand for island home sites. Cortes is watching the effect of development and tourism on Quadra. A politician's remark that the Gulf and Discovery Islands were to be "the recreation grounds of Victoria and Vancouver" was received with little enthusiasm by islanders. One hopes that the provincial parklands will be carefully supervised, as they have been on Rebecca Spit. Gorge Harbour could develop pollution problems with its narrow-necked, land-locked harbour if there is unwise sale of shoreline property. The stated policies of Cortes and the Regional Board, however, seem to indicate that both board and residents are anxious to guard the natural environment and preserve the independence of the island.

CHAPTER 3

Read Island

Partly because of its wild, rugged terrain, its isolation and its lack of law enforcement officers, Read Island's early history involved murders, missing men, and skeletons found in shallow graves. A lonely island, dwarfed by its larger neighbours Quadra and Cortes, without scheduled ferry service for easy access, Read Island managed nevertheless to attract a number of settlers before 1900.

Alex Russell, logger, who pre-empted and received Crown Grants on several of the islands, acquired over 600 acres of land on Read in 1883. Mallandaine's *British Columbia Directory* for 1887 lists Joseph Silva, farmer, under "Reid [*sic*] Island." The 1892 directory notes: "Read Island, lying to the east of Valdes Island, has several pretty farms and a store." It lists three names on Read: Peter Schuffer, farmer, Richard Davis, farmer, and F. Willmot, storekeeper. Davis pre-empted land in 1892 as did John H. Smith, farmer.

"A REIGN OF TERROR!" Under this blaring headline

the usually staid Victoria *Daily Colonist* reported on July 5, 1893: "In the Read Island murder case British Columbia has developed its first real desperado of the Jesse James variety.... This man, armed to the teeth, has for more than an entire week bade defiance to the law..."

Jack Myers, alias Bart McKenzie, alias Ben Kennedy, had come to Read Island in his sloop on June 24, bringing with him 60 cases of whisky, stolen from the Evans, Coleman and Evans wharf in Vancouver and sold to him for $120. Armed with two Colt revolvers and a hunting knife, and accompanied by his dog, Kennedy made his way to a branch of Taylor's logging camp that was working at White Rock Bay in the northwest area of the island. Salem Hinckly was in charge of the camp, his crew being Angus Cameron, John O'Neil, Jim Burns and Jack O'Connor. For $2 a bottle, Kennedy offered the whisky that had cost him $2 a case. The loggers bought $52 worth of the liquor and began a weekend of celebration.

By Sunday night no one was sober. Around one o'clock Monday morning, Kennedy began to boast of the obedience of his dog, which he said would prevent anyone from touching its master's property. He threw his waistcoat on the floor and commanded his dog to guard it. O'Connor made a sudden dive for the waistcoat, the startled dog scuttled under a bunk, and Kennedy in a rage drew his gun and fired at the animal as it fled from the cabin. When O'Connor protested, Kennedy covered him with his gun. O'Connor tried to reach his rifle, knocking over the lamp, and the two men grappled on the floor. O'Connor shouted for help to Hinckly who was outside. A shot was fired, and Hinckly ran in to find O'Connor fatally wounded.

Kennedy prowled about the camp, threatening to shoot the men unless they backed his claim of accidental death. Hinckly, who was fairly sober, managed to slip away to a boat and inform Michael Manson, the magistrate on Cortes, who came back with Hinckly on Tuesday. Manson examined the body while Kennedy stood by, armed with a 44 Winchester and two revolvers in his belt. Manson and his deckhand were unarmed, and the loggers' rifles were unloaded, so Manson, displaying his habitual calm, took down Kennedy's statement, ignoring his boasts that he had been one of Jesse James' gang, had robbed a stage in Deadwood City, had shot a man in a bar-room brawl and was wanted in the United States. Since Kennedy swore he would kill Manson if he tried to arrest him, the magistrate assured him he need have no worries if he was innocent, and told the loggers to bury the dead man.

Later, an autopsy was ordered, and the body was disinterred. Dr. Walkem, the coroner, rode on horseback through the bush from Nanaimo to Comox to hold an enquiry, with Manson and the loggers as witnesses. A warrant was issued for Kennedy's arrest.

For the newspapers, the case had all the aspects of a Wild West thriller. A "posse" left Nanaimo headed for Read Island. A $500 reward was offered for Kennedy's arrest. It was rumoured that the outlaw had hidden his sloop and taken refuge in the woods on the mountainside. Constables Anderson, McKinnon and McLeod, with Chief Stewart, left Comox on the *Estelle* to conduct a search, Superintendent Hussey following on the *Joan*. Seventeen men in all took part in the search, first up Ramsay Arm, then up Bute Inlet where campfire smoke was reported on the mountainside. They found the campsite, with venison roasting over the fire, and surrounded Kennedy who was hiding among the trees. Kennedy, thin and half-starved, surrendered when he learned that his venison had been confiscated. A week earlier he had eaten his dog.

The outlaw was brought down to New Westminster for

trial. Though he admitted to forgery, stealing logs and smuggling whisky, he swore O'Connor was shot when both were struggling for the gun. The verdict was manslaughter and the sentence life imprisonment.

An editorial in the *Daily Colonist* on July 14 entitled "The Law Supreme" stated: "The arm of the law in British Columbia is both strong and far reaching." It congratulated the police and said they had made Read Island once more a safe place for its residents.

Lack of daily transportation to and from the other islands has always been the isolating problem of Read, but steamship contact with the outside began almost at once for the settlers when the *Comox* began its sailings to northern logging camps and settlements in 1892. Read Island was a twice-weekly port of call, and Evans Bay is mentioned as a port of call in 1894.

The first post office was established in 1893, and Edgar Wylie is listed in the directory as its postmaster. Wylie. came from New York, built "Wylie's Hotel" in Burdwood Bay and acted as trader, merchant and hotel-keeper until his death in 1908.

Alan Greene wrote: "Strange stories of men who disappeared seem to gather round Wiley's [*sic*] name and his rather notorious hotel."[1] The hotel must have seen some rough nights: Bill Whittington, formerly of Read, says he and his brother-in-law, Walter Maclean, who purchased the property, found numerous bullets imbedded in the hotel walls when they were renovating the old building as a house for Maclean and his partner, Murray.

The Wylies were witnesses in a murder trial which *The Province* of November 23, 1895 called "a particularly sensational one." In October of 1894, a body was discovered lying in the bottom of a skiff adrift in Sutil Channel. Two men in a sloop, headed for a logging camp, discovered the body and towed the skiff to the nearest settlement, Burdwood Bay on Read Island, where Wylie had his store and post office. The Wylies identified the dead man as Chris Benson, farmer and logger, and a partner of Ed Wylie in his store. John Smith, a neighbour, suggested the dead man might have suffered a heart attack and fallen against the side of the boat, accounting for the injury to the side of his face.

The sloop brought over the nearest Justice of the Peace, Michael Manson of Cortes. Necessity had turned Mike into an amateur detective, and he settled down to examine the corpse. He discounted Smith's suggestion that Benson's injuries, which included a broken nose, were due to collapse in the boat. He noticed too that despite the severe injuries there was no sign of blood on the skiff. He took statements, boxed the corpse in a rough coffin and took it with him by steamer to Vancouver. There, Dr. Bell-Irving, the coroner, performed an autopsy.

The mystery deepened with the doctor's report: Benson had been struck repeatedly on the head and cheek with a blunt instrument. All the head wounds had been inflicted before death. The lungs were free of water; both heart and lungs were healthy.

Chief Inspector Hussey of the provincial police met with Manson in Comox to discuss the case. Who had killed Chris Benson? Manson discounted Indians, for liquor was usually involved in cases of Indian violence, and Benson did not drink. He felt that Ed Wylie had been genuinely surprised and shocked when informed of the death; for that matter, Benson and the Wylies and John Smith and his wife had all known one another in the United States and had come up to Read from the middle west the year before. Both couples claimed to be on friendly terms with Benson, who had a wife and two daughters in San Francisco. Hussey suggested that money, liquor or women were the usual reasons behind violence. None of them possessed enough money to warrant murder. None of them drank heavily. There

were only two women: Mrs. Wylie and Mrs. Smith. Mrs.
Wylie was a quiet, reserved woman who taught three of
the Smith children in a room behind the store; Mrs.
Smith had four children, which seemed to make her in-
volvement in a killing unlikely, but she was a talkative
woman and more apt to reveal information if any was
being concealed.

Manson conceived a plan to send a local fisherman, Bill
Belding, as an undercover man to encourage Mrs. Smith
to talk. Belding soon won her confidence, and she con-
fessed to him that her husband had murdered Benson.
Then Hussey interviewed her and she told him the
details. Benson, she said, had been visiting her secretly
when her husband was away; washing on the clothes-line
by day, or a light in the window by night were signals
that the coast was clear. One morning Smith left for a
hunting trip and Benson, returning from a business trip
to Cortes, saw washing on the Smith clothes-line and
pulled his boat into the cove below their house. The
hunting trip was cancelled and Smith returned unex-
pectedly to find Benson embracing his wife. Enraged, he
seized a wooden mallet and struck Benson again and
again. The two Smiths carried the body to the skiff, and
John Smith towed the skiff behind his own boat and set it
adrift half a mile down the channel. The couple scrubbed
blood from the floor, and according to Laura Smith, Mrs.
Wylie helped with the cleaning up. Laura signed her
statement for Hussey, who procured a warrant and returned to arrest Smith.

The trial was held in Vancouver at the fall assizes of
1895 before The Honourable Mr. Justice Walkem. The
Crown produced planking from the Smith house which
an analyst swore had been bloodstained and scrubbed
clean. William J. Bowser, who later became premier of
the province, was the defence lawyer, and he called Ed
Wylie who swore the blood was caused by the butchering

of a deer. Despite the evidence against Smith, which included his children's testimony of the sounds of blows and groans from the bedroom, the jury found him not guilty.

The Province of November 23 commented: "An incident occurred which raised the anger of the presiding judge. After the jury had been dismissed the discharged prisoner stepped down from the box and most improperly proceeded to shake hands with the jury. This was altogether too much for the judge, who instantly put a stop to such suggestive antics and ordered the offender to leave the court. One does not like to see such a scene in a court of justice; a frequent occurrence of it might give rise to the suspicion that more than good-will had passed between the jury and the accused."

Ed Wylie was buried in a crack in a huge boulder on Lot 340 on a point by Bird Cove, known locally as Healey's Point. He was a small man, but the crack failed to conceal him entirely, so cement was poured into the crevice. Before it hardened, small stones were set into the cement to spell out ED WYLIE and the date of death. The grave was seen there for many years but eventually roots of a nearby tree split the cement and the skeleton was revealed, perhaps giving rise to some of the stories of "skeletons found in shallow graves."

A happier record remains of the children of Harley Wylie, farmer, in the Public Schools Report of 1895. This states that the first school on Read Island was opened in October of 1894 at Evans Bay, the teacher being a Miss E.M. Carter at a salary of $50 a month, with ten pupils, four boys and six girls. It continues with a list of prizes (commonly published in the Public Schools Report in those days) and the winner of the award for Deportment for 1894 is Harley Wylie, while Minnie Wylie is winner of the prize for Regularity and Punctuality.

John Jones, a Welshman, who pre-empted 160 acres in 1894, added another 400 acres a few years later. Thirty years after this, making his living by farming, fishing and logging, Jones had a fine orchard and garden, a singular house situated half a mile back from the sea and the job of postmaster at Surge Narrows. Jones' first house was burned out, so he rebuilt it to be as fireproof as possible. The site was on solid rock, the cement foundation many feet thick, the basement blasted out of the rock, and the outside of the house sheeted over with tin. Jones' post office was a tiny shack. He was overly conscientious about his postal duties; hours ahead of boat-time he would row the half-mile out to the landing and sit in the shed with his feet wrapped in sacking to keep out the cold on winter nights. Forty-eight years after his pre-emption date he finally abandoned the property he had toiled upon for so many years and he and his wife were taken to St. Michael's Hospital at Rock Bay.

Another old-timer was big Charlie Rosen who worked on the road gang. His house fronted a lake in the southern part of Read, and his fine barn was composed of hand-hewn timbers with a cedar shake roof. Charlie was devoted to his horse, which was named Lindy after Colonel Lindbergh. One fateful day in the orchard, Lindy spied a tempting apple just out of his reach, leapt up, missed the apple and came down with his head and neck caught in a crotch of the tree. There Charlie found him hanging, dead. Grief-stricken, Charlie's thrift nevertheless overcame sentiment, and he cut off roasts and steaks from Lindy before digging a pit by the tree, sawing through the restraining limbs, and letting his horse drop neatly into the grave. Charlie's neighbour Murray passed by the road gang some time later and saw Charlie eating his lunch in a shady spot.

"Did you hear what happened to Lindy?" Charlie asked him, tears in his eyes. "He hanged himself!" Sobs choked him as he held up a thick slice of smoked meat. "Here is Lindy!" he said, his face a mask of tragedy.

Lumberman Moses Ireland was a familiar figure on all
the northern Gulf Islands in the 1880's and was featured
in various newspaper articles and history books. He lived
at different times on Read Island, Cortes Island, Camp
Island (a group of islands later named the Subtle Islands)
and at Bold Point on Quadra. Like many of the early
settlers, Ireland was born in the United States. He left
school at 14 to work in a lumber camp, made small fortunes in the California and Cariboo gold rushes, and
started a mill in Moodyville that failed when he lost
money on his shipments. A romantic story often told of
him is his rescue of a party of 30 travellers lost in a snowstorm on Bald Mountain. Ireland led the main party to
shelter in a cabin, carrying a 14-year-old girl on his back,
then located several men who had wandered off in search
of help and rescued them as well. Judge Howay's history
of British Columbia praises Ireland for refusing a
proffered reward. Some of Ireland's stories were taken
with a grain of salt by islanders, as he had a vivid imagination and a droll sense of humour. Cortes Island
children suspected of stretching the truth were
admonished by parents: "That sounds like a Moses
Ireland story!"

In 1880 Ireland recognized the profit to be made in
buying up land and holding it for a higher price. Land,
timbered or untimbered, was $1.00 an acre in British
Columbia, and when Ireland heard a rumour that the
price of timberland was to be raised to $2.50 he got large
logging firms such as Sayward and Merrill to advance
him money, and then sold the land advantageously to the
lumbermen. After 1887 the Land Ordinance Act put a
stop to this easy money-making, requiring applicants to
declare timberland and imposing a royalty on timber cut.
A licence was also required, and four years later land had
to be surveyed and classified before purchase.[2]

In 1905, Moses Ireland told *The Province* he had seen a

mysterious floating light up the coast, like a woman in white. "I think it's a warning," he said. According to Justice of the Peace Bagot, Ireland's death in 1913 on Bold Point was an unsolved murder mystery.

Henry Tiber, who farmed on central Cortes in the days of sailing ships, also lived on Read Island. Tiber was a boatbuilder and, for a workshop, constructed a huge shed of split cedar, 60 by 20 feet, that was a landmark on Read for many years. He built several clinker boats of excellent workmanship, much in demand and known as Tiber boats.

Tiber sold out to William Brockman who came from Oregon, tried his luck in the goldfields, and then worked at logging on Read. Logging by oxen was at an end, and loggers used horses or, in Brockman's case, a steam donkey. Brockman is listed as postmaster in 1922. His son-in-law logged on Read and his grandson Bill Whittington was born there, married a Read Island girl and lived and logged at Evans Bay. Three generations of the Brockman-Whittington family lived on Read. Cougars were prolific on the island during the 1920's and 1930's. Bill Whittington's diminutive sister Bonnie lost her dog to a cougar and vowed to rid the island of the predators. She shot ten of them.

Before there were stores on the island, supplies were brought in by rowboat from Whaletown on Cortes. Most settlers used oars; few owned engines. The elder Whittingtons managed the post office and mail collection at Evans Bay before the store was built. There was no wharf then, only a float anchored well out from shore so the Union Steamship vessel could get to it at either high or low tide. There was no set time for boat arrivals as the amount of freight to be loaded or unloaded was the deciding factor. Whittington was away from home in logging camps much of the time, and the task of rowing the heavy bags of mail out to the float, often in rough seas and in the blackness of night, fell to his wife.

Alan Greene recollects the difficulties experienced in moving patients down the steep cliffs of Read Island to the Columbia Coast Mission's hospital ship. In one such incident, a woman who was a stretcher case was being lowered in a rowboat down the little logging railway. Her husband was working the winch and suddenly lost control of it; the rowboat launched itself, was caught by the swift current and headed out to sea. The men ran with frantic haste to find another boat and row in pursuit, finally overtaking the rowboat as it was gathering great speed in the rapids.

Over the years, there is no mention of successful mining on Read Island in the Annual Report of the Minister of Mines. Bancroft's geological report on the islands tells of post-glacial, marine clays near the shores of some of Read's bays. Bancroft found the clays bluish in colour, compact in texture and possessing a tenacious habit when wet, and he adds: "Mr. E.W. Wylie of Burdwood Bay, Reade [sic] Island, informed the writer that he used such clay, taken from a deposit immediately back of his house, for lining his cooking stove and that it proved to be satisfactory in every way."[3]

The unspectacular history of the Solyman and Freya claims, staked in 1920 and 1921, is followed in the Mining Reports for some 25 years. The 1920 report says the two claims are owned by H.N. Bacon of Quathiaski Cove, and situated on the west shore of Read, but adds that samples proved the gold values to be very low. In 1938 the report also mentions the Shackles and the Anona, "variously owned by Chas. Tweedie of Fanny Bay, and C.A. Carlson and associates of Quadra Island."

"Lord" Bacon, as he was always called by the islanders, was certain that a fortune would eventually be gleaned from the Solyman and Freya. He arranged for Alan Greene to draw up a will leaving the expected bonanza for the erection and maintenance of a home for children, a superlative home that would provide a good life for "happy children." Bacon died in his late 70's, a poor man, his dream of happy children unfulfilled.

The problem faced by all remote communities, of finding sufficient pupils to reach the quota required for a school, was a constant anxiety for Read Island. After the early start in 1894, when Evans Bay on the east side had the first school for white children among any of the neighbouring islands, there was a long gap from 1905 through 1921 with no mention of Read in the list of assisted schools in the Public Schools Report. It is listed again for the 1921 to 1922 term, with Miss M. Essler as teacher. Classes were held in Wylie's Hotel in Burdwood Bay until a schoolhouse could be readied.

On the west shore of the island, Robert Tipton, farmer, storekeeper and Justice of the Peace, had a strong personal interest in obtaining a school for Surge Narrows. He and his wife Ethel had adopted their two nephews and niece, the orphaned Keeling children. In 1925 they had engaged a teacher, Maudie Pierce, to instruct the three in a tiny shack on the beach in front of their home. Postmaster John Jones promised a piece of his land for a school if a quota of children could be rounded up. Tipton sought a solution by inserting a notice in *The Province*, offering to locate pre-emptions for desirable settlers with children.

The 35 replies that Tipton received from all over British Columbia and Alberta contain tragic tales of past failures in city or on farm, of the auctioning of livestock and equipment for next to nothing, and of pathetic hopes for success in a new environment. Some lost interest after learning of inconveniences such as the need to row supplies a long distance across the water, but one man declared that after hearing Tipton's description of the island scenery, he would be content to suffer poverty and hunger in such beautiful surroundings, for he could "feast his eyes."

One of the letters to Tipton was from Charles Redford, who had homesteaded in Alberta and run a prosperous farm there for 15 years. An unlucky move to Manitoba resulted in two successive crop failures. He was trying his luck in Lone Butte in the Cariboo district in British Columbia when he saw Tipton's advertisement. On July 29, 1925, Redford wrote to Tipton: "We have decided to come to Surge Narrows in the near future."[4]

Redford brought his wife and two young sons to Read in the fall of 1925 and settled near Surge Narrows, where he helped the Tiptons to build a log-cabin school. When the Tipton horse strained a shoulder hauling the logs, the Tiptons' two bulls were harnessed and persuaded to take over the task. Mabel Plommer was the first teacher in the log school, 1927-28. The eldest Redford boy, Harold, adapted at once to pioneer life by the sea. Fascinated by boats, he built his first dugout canoe while he was a teenager, graduated to rowboats, and finally to sturdy gasboats complete with cabin. With the tenacity of a number of pupils in these outlying islands, he earned a high school diploma, taking his courses by correspondence and doing his studying in the little log-cabin school. The proximity of Read to Maurelle led to romance. Young Redford often paddled his dugout to Maurelle to visit the Armstrong family who had come to Maurelle from California in 1928. Eventually he married his boyhood sweetheart, Ethel Armstrong.

In 1926, Doris Nye, a 19-year-old school teacher, spent a year teaching on Read Island. The Union Steamship *Catala* landed her at four in the morning at Bold Point on Quadra where she was met by young Bob Bell and taken to his parents' farm for a hearty meal to fortify her for further travel. (The Bell farm was originally the cattle ranch of Moses Ireland who was buried on the land after his violent death in 1913; Ward and Elder were logging there when the Bells took over. The Bells ran the post office at Bold Point as well as managing their farm and cattle ranch. Since the steamship service has ceased, Bold Point has been virtually abandoned.)

Carl Ekman, secretary of the school board, ferried the young teacher across to Surge Narrows in a leaky rowboat, its sides draped with moss and seaweed. He stood up facing the bow, rowing with quick, choppy strokes in the island manner. As water seeped through the boards, he instructed his apprehensive passenger to bail out the boat with a tobacco tin and to ignore an approaching pod of killer whales. Nonchalance was difficult, as each of the great mammals was many times the size of the tippy little rowboat, but the killer whales, belying their name, passed peacefully under and around it.

The school was on the opposite side of the island, in Burdwood Bay, and is described by the teacher (now Doris Davies): "This had been a cookhouse in its heyday in a logging camp, and had been floated down the coast and set up on the rocky beach. It served its purpose admirably for the 13 pupils, grades two to seven. Three planks formed the blackboard painted with some green substance, and a pot-bellied stove provided roast or chill, depending where you were. . . . The playground was a small rocky beach, where the children amused themselves. No organized activities were needed or expected— just imagination and inventiveness. Some of the boys made and operated a log-hauling apparatus, complete with donkey and high spar."

Of the island residents, she says: "I found the local settlers of great variety and much independence. They, for the most part, operated small 'stump ranches' under great difficulty, and the rare occasions when they saw cash were usually for a few days' work on the roads. Properties were scattered and boats were few." It was not all work and no play. Young Pete Lilja had a boat with an engine and took Doris to a dance at Heriot Bay. "Never will I forget that hall, with the thundering boots of those

Scandinavian loggers crashing around in polkas and their own versions of hilarity. The music was loud and infectious, but scarcely could be heard with the din of voices and stomping."[5]

Doris Nye's position as teacher, paying $90 a month with $35 deducted for board at the Ekman home, lasted only the one year, as a family with school children moved away, lowering attendance to below the required quota. By 1928 there were schools on both the east side and at Surge Narrows on the western shore.

Read Island's largest population was in the late 1920's and early 1930's, reaching roughly 100 residents. There were two stores, Lambert's in the southeast and Tipton's in the northwest. The island was a comparatively busy spot then, with logging at Evans Bay, and the Union Steamships and Columbia Coast Mission boat calling regularly. In the 1940's there were four logging companies operating, and a fish-buying station served the local fishermen.

Several farms on Read were severely damaged by the earthquake of 1946. On the Marshall farm, a ten-acre orchard and wheat field were completely destroyed. Huge sections of gently-sloping, fertile farm land dropped away, leaving deep chasms filled with sticky blue clay. Fruit trees dropped into holes and disappeared from sight, while others were left standing on the brink of the chasms with their roots bared of earth. Many fissures, eight to ten feet deep, were also formed, filled with water and blue clay like the chasms. In some sections of Read Island, geysers of blue clay coated stumps and trees, while in other parts the clay rose slowly under great pressure and then gradually receded, leaving the ground saturated with the blue mixture. This seems to be the same blue clay of which Bancroft spoke in his geological report, which was used by hotel-man Wylie to line his cookstove in the first years of the century.

In the 1960's, the population dwindled to about five

families. The two schools closed, logging ceased, and most of those who remained commuted from Read to other islands to work. The Union Steamship Company had sold out and no longer served the islands.

All the old-timers are gone now from Read, either dead or moved to islands offering medical care and easier living conditions. There are still farms in the centre of the island and around Surge Narrows. Some residents live on Read and log on other islands; others rent their homes to summer vacationers. Several city people have bought land for summer homes. A new population of young people began to move in during the 1970's, some on welfare, squatting on the land, others buying property and working either off the island or on it at such trades as boatbuilding. A few have established their own power plants to supply electricity on their property.

The small, log-cabin school at Surge Narrows was replaced in 1949 by a Quonset hut that serves nearby islands without schools, as well as Read. The teacherage house was flown in by helicopter, dangling from the clouds at the end of a long rope, as this was found to be less expensive than shipment by tug and barge. Surge Narrows has a government float, a store and a post office. Since there is still no ferry connection with the other islands, and no Union Steamship mail service, the mail is now flown in by plane. There is no inter-islands telephone service, but a local system serves the residents on Read.

It is interesting to note that most of the Discovery Islands bear Spanish names:—Quadra, Maurelle, Sonora, Redonda, Cortes, Hernando and Marina,—but Read, in the very centre of this Spanish group, is uniquely British. The island, its bays adjoining Hoskyn Channel and Drew Pass, and the small islands within them all bear names chosen by Captain Pender of the Royal Navy to honour Britishers in the Admiralty's Hydrographical Department.

Read Island's name honours Captain William Viner Read of that department, as does Viner Point, the popular fishing area on the southern tip of the island. Mayes Point, at its northern tip, Evans Bay, Hoskyn Channel, Dunsterville Islet and the King Islets in Hoskyn Channel and the Penn Islands in Sutil Channel are also named after members of the Hydrographical Department. Drew Pass, after Charles Randolph Drew, R.N., and little Hill Island, just outside the entrance to Evans Bay (small in size but reaching 490 feet in altitude), after survey officer Captain Stephen Hill, R.N., were also named by Pender while he was surveying with the *Beaver*.

CHAPTER 4

The Redonda Islands

The Spanish captains, Galiano and Valdes, named the Redonda Islands in 1792. Their chart marks the spot *Isla Redonda* (Spanish for "round island"), the two islands together forming a rough circle and drawn as one island by both the Spanish and English explorers.

The explorers of both nations charted the waters around the Redondas, and the English made Teakerne Arm, on West Redonda, one of their anchorages. The English party also noted a lofty mountain, Mount Addenbroke, on East Redonda, using it as a landmark to remind themselves of the deserted Indian village on the mainland bank opposite, where Vancouver's men were attacked by "myriads of fleas."

In 1792, Spanish and English were simultaneously exploring the islands and waterways of the inside passage. Earlier, in 1790, before the arrival of Vancouver, Francisco Eliza, commandant at Nootka, had despatched Manuel Quimper in the *Princesa Real* to explore and map the Juan de Fuca waters, and Georgia Strait had been mapped by the Spanish almost to the north end of Texada Island. When Vancouver's magnificent volumes of the English expeditions were published in 1798, the Spanish realized that both their earlier and their simultaneous explorations should have been more widely publicized, but the Spanish government at this time was absorbed with political problems and only a brief general outline was published. In 1802 a more detailed account was brought out by the printer Navarrete and translated into English by Barwick in 1911; it was too late, however, to win the interest and acclaim that had been accorded Vancouver's account.

Captain Vancouver anchored his two ships in Lewis Channel by West Redonda in June of 1792 and sent out survey parties to explore the mainland coast, looking always for a northwest passage. The lonely, forested islands had little appeal for Vancouver, who named the waters around the Redondas "Desolation Sound" and commented: "Our residence here was truly forlorn; an awful silence pervaded the gloomy forests, whilst animated nature seemed to have deserted the neighbouring country, whose soil afforded only a few small onions, some samphire and here and there bushes bearing a scanty crop of indifferent berries. Nor was the sea more favourable to our wants, the steep rocky shores prevented the use of the seine, and not a fish at the bottom could be tempted to take the hook."[1]

Vancouver's map names the whole of the Redonda waters Desolation Sound but only the entrance now bears the name. Lieutenant Broughton of Vancouver's party also found the shores "inhospitable and forlorn" but enjoyed the "excellent spruce beer" the men brewed there to ward off scurvy. Menzies, botanist of the expedition, described the land with more enthusiasm, admiring the beautiful contrast of maples with the "gigantic aspect &

dark verdurous hue of a thick forest of Pinery." The unfavourable reaction of his survey crew was mainly due to the unhappy experience in a nearby cove where, Menzies says, "the picturesque ruins of a deserted Village placd on the summit of an elevated projecting Rock excited our curiosity & inducd us to land close to it to view its structure."

Menzies said of the disastrous encounter at "Flea Village": "The narrow Lanes between the Houses were full of filth N nastiness & swarmd with myriads of *Fleas* which fixd themselves on our Shoes Stockings & cloths in such incredible number that the whole party was obligd to quit the rock in great precipitation, leaving the remainder of the Assailants in full possession of their Garrison without the least desire of facing again such troublesome enemy. We no sooner got to the Water side than some immediately stripped themselves quite naked & immersed their Cloth, others plungd themselves wholly into the Sea in expectation of drowning their adherents, but to little or no purpose, for after being submerged for some time they leapd about as frisky as ever; in short we towd some of the Cloths astern of the Boats, but nothing would clear them of this Vermin till in the evening we steepd them in boiling water."[2]

Galiano and Valdes had also anchored their ships in Desolation Sound and sent out their longboats to explore the channels and rapids that lay ahead. Galiano went with the *Mexicana's* longboat to explore to the south of Desolation Sound, around Kinghorn Island, and named Malaspina Inlet *(Brazo de Malaspina)* after Alejandro Malaspina. (This is the present Okeover Inlet; the entrance, which he named *Canal de Dos Bahias* is now known as Malaspina Inlet.) It was Malaspina who had suggested that Galiano and Valdes be given command of the *Sutil* and *Mexicana* to explore the northwest coast, when illness prevented the officer originally delegated for the voyage from undertaking the commission.

The English met Valdes in his launch and were told that he had just completed the exploration of Toba Inlet (*Brazo de Toba.*) This name is a Spanish mapmaker's error. Valdes called it *Tabla*, or "board," after a painted Indian board found on the shores of the inlet. Valdes showed his survey chart to the English, proving the inlet terminated in shallow water after about 11 miles, but they decided to examine the waters themselves.

It has been suggested that this decision offended the Spanish captains and was the cause of the separation of the two parties in Lewis Channel, but in fact, both parties usually made independent surveys. It seems likely that the Spanish kept to the mainland coast due to the specific orders that had been given to Maurelle, the explorer who became ill and whose commission was then given to Galiano and Valdes. Certainly the most friendly relations were recorded on both sides. Charts were exchanged, and Galiano, who had visited Nootka in early June, gave Vancouver a letter of introduction to Quadra. In the diary that Galiano kept for the viceroy, he wrote: "The English frequently assured us that they were sorry to see us go that way as the navigation was very dangerous. Nevertheless we undertook it as we thought it necessary. This obliged us to separate after an intercourse, not only harmonious, but of the closest friendship. We dined together various times, sometimes in their vessels and sometimes in ours."[3]

Vancouver seemed pleased to be leaving Desolation Sound. As the *Discovery* crossed the gulf, he noted: " . . . the scene was more congenial to our minds, not only from the different aspect of the shores, but from the attention of the friendly Indians, who, as we were crossing the gulph, visited us in several canoes, with young birds, mostly sea fowl, fish, and some berries, to barter for our trinkets and other commodities."[4] The *Chatham* followed the *Discovery* after a delay caused by the final unpleasantness which they associated with Desolation Sound.

The *Chatham*, which the Spanish referred to as "very ill-shaped", hooked her anchor on a rock and "every means that could be devised had been resorted to without effect, until the moment they were about to cut it away it cleared itself, which fortunately saved the anchor and cable."[5]

Vancouver had been amazed at the Spaniards' temerity in attempting their exploration in the small, ill-equipped vessels *Sutil* and *Mexicana*. Their misgivings were justified, for the Spanish captains immediately ran into difficulties with the tides and currents of Lewis Channel. Finally, they furled their sails and rowed the ships close to shore to avoid the force of the rapids.

Captain Richards, surveying in 1862 in the navy's vessel *Hecate*, named Waddington Channel, which separates the two Redonda Islands. He named it after Alfred Waddington, an early pioneer of British Columbia who came to Victoria from England in 1849 and spent his entire fortune attempting to establish a wagon road from the head of Bute Inlet to Fort Alexandria and eastward. On April 30, 1864, Chilcotin Indians massacred 14 of his road crew of 17 men and destroyed his tools and provisions, forcing Waddington to abandon his cherished dream.

Captain Pender, surveying in the *Beaver*, named Homfray Channel, east of Redonda, after Robert Homfray, a civil engineer living in Victoria. Walbran says of him: "He was of an eccentric disposition, and for some time before his death had his tombstone erected in Ross Bay cemetery, with all particulars on it with the exception of the date of his decease, which was added after the 19 September, 1902, the day he died."[6] Walbran also says that Connis Point on the northern tip of West Redonda was named after Captain Pender's Skye Terrier Connis which travelled with his master in the *Beaver*. Pender and Richards are given joint credit for the naming of Lewis

Channel, west of Redonda, after Captain Herbert George Lewis, an officer in the service of the Hudson's Bay Company. Lewis was first officer of the *Otter*, which accompanied the *Beaver* when she navigated the inside passage, Lewis acting as pilot and interpreter.

The Redonda iron mine, Elsie, was located in 1892. De Wolf and Monro of Vancouver received Crown Grants for the land on the north shore of West Redonda. One year after the location, 626 tons of magnetic iron were shipped to an Oregon smelter, but shipments ceased after this, and from 1901 no development work was carried out for several years, except for some quarrying of limestone and laying of trails. In 1920, 8,450 tons of limestone rock were shipped to the Whalen pulp and paper mills. No evidence remained by then of the cabin or chute of the Elsie iron mine, and during the 1920's there was a slump in mining everywhere in the province.

A beautiful pink granite in Walsh Cove on the northeastern shore of West Redonda was said by Bancroft in 1913 to be somewhat similar to the celebrated Baveno granite from the vicinity of Lake Maggiore, in Italy, "but the shade of pink is more delicate and its general appearance more pleasing."[7] The geology of the island suggests large deposits are possible on West Redonda, though none are known to exist.

On East Redonda, a small area of arable land on the east coast attracted farmers, who were successful in raising vegetables and apples, and hogs for market and domestic use. By 1918 a survey report to the Minister of Lands showed that settlers on Refuge Lagoon in the south of West Redonda were making good progress. Lot 4936 was singled out as having a good house and garden on 15 acres of land on the shore of the lagoon. At the head of Teakerne Arm, settlers were also getting excellent vegetables and fruit crops; fur farming was carried on at Refuge Cove, with hand logging the usual supplementary occupation. There was a dance hall at the rear of the Refuge Cove store in the 1920's, and settlers from miles around gathered there on Saturday nights. Redonda Bay, first mentioned in the 1922 directory, is listed with a shingle mill, cannery, general store, and a post office run by Norman Dillabough. Cougars were an early problem on the Redondas, and increased again in the 1960's, killing chickens, dogs and deer.

The first school on the Redondas mentioned in the Public Schools Report is in 1914-15, when Refuge Cove acquired a government-assisted school. The teacher was Miss G.H. McGregor with a salary of $60 a month, and trustees were Mrs. B.M. Black, secretary, Mrs. A. Hoyel and Mrs. S. Tosr. There are listings in the early 1920's, then no more during the 1930's and 1940's. In the 1949-50 term the school appeared again, serving nine pupils, after which there is another lapse.

Although for the most part the Redonda Islands have had a shifting population, some individuals were residents for many years, notably storekeepers, whose trade was more stable than the industries which came and went. Syd Vicary and his daughter had the store and post office at Redonda Bay during the 1920's and through the Depression years. Norman Hope and his wife had lived 23 years at Refuge Cove before their store burned down in 1968. Undeterred, they converted an 87-foot barge into a store, stocked it with groceries, hardware and boat fittings, and docked it at the government wharf. Though the Hopes have retired, the store is now a co-operative enterprise which provides all facilities required by holiday boaters. A scheduled plane service has just been started, between Refuge Cove and Vancouver. Resident population fluctuations have followed the rise and fall of the industries; in 1931 the population of Refuge Cove was 200, according to the *British Columbia Directory*, whereas the *Postal Householders' Directory* gives the 1975 population as 21. However, the large number of summer visitors is apparent when boats jam the floats in the sheltered cove during July and August.

Thomas Manson compiled a history of Redonda Bay from 1835 to 1975 in a manuscript available in typescript in the Provincial Archives. In 1835, he says, there were two Salish Indian villages with populations of about 100 each on West Redonda, and a summer village near Deceit Bay. Remains of a stone circle, a type of weir they used for trapping fish, can be seen between tides at the mouth of Lillian Russell Creek. Other traces of the Salish, long gone from the district, are petroglyphs on the east side of Ellis Lake.

Hand loggers arrived on the island during the last quarter of the nineteenth century, traces of their work areas showing in old log chutes and the remains of oxen skid roads. The Redonda terrain was ideal for hand logging; the steep cliffs that ran down to the sea made it easy to get logs from forest to water, and the many surrounding islands sheltered the booms as they were towed away. Getting supplies to the Redonda camps in those early days must have been difficult, however, for there was no steamboat service to the Redondas until the 1920's, nor did the Columbia Coast Mission boats make regular calls to them.

After 1920 the larger logging companies moved in, among them Mosher Brothers Logging Company, with mechanized equipment. Around Redonda Bay remains of the "rail show" can. be seen, though Manson says "the rather grandiosely named 'railroad' was probably just a wooden track along which an old steam 'donkey' pulled a wheeled vehicle by means of a cable, or simply dragged the logs on skids."[8]

It was a cannery that was the nucleus of an emerging settlement. Mr. Francis Millerd suggested this was the Redonda Canning Company, established in the early

1920's. (The 1918 Mines Report mentions Deceit Bay
Cannery, and a photograph of it in the Provincial
Archives is dated circa 1915. The 1922 *British Columbia
Directory* lists Redonda Canning and Cold Storage
Company.)

What Manson calls "possibly the most exotic thing ever
to appear at the settlement" was the former ferry S.S.
Transfer, a vessel which had been used on a run from
New Westminster to Chilliwack; now that she was no
longer fit for public service, she had been sent to Redonda
Bay to serve as a boiler plant for the cannery. For a number of years the old paddle-wheeler lay on the beach,
supplying steam for the Redonda Canning Company's
machinery and occasionally providing shelter for some of
the workers.

Regular steamship service came to Redonda Bay in the
early 1920's when Union Steamships *Comox*, *Cassiar*,
Cowichan and *Cheakamus* all called there. Before
Stuart Island was developed, Redonda Bay was a major
stopover point for ships to wait for the best tides before
attempting the rapids. To serve these transients and the
local settlers, a general store was built.

Then came the Depression years and the cannery, the
cold storage plant, the logging operations and the shingle
mill all closed down. Only the Vicarys' store remained
open.

The early 1940's saw an economic revival when Francis
Millerd and Company opened the Redonda Bay cannery
and reduction plant with George Henshall as manager.
Old *Transfer*, the paddle-wheeler, was dismantled and
her boilers brought up from the beach and moved inside
the cannery. A dam was constructed above the bay to ensure a year-round supply of water for electricity, since
Lillian Russell Creek was apt to run dry.

In 1938, Gerry Olmstead, a logger, came to Redonda

Bay with his wife and four sons. Until cabins were built, they camped on the beach in a tent composed of a wooden frame covered with potato sacks and blankets, but in a short time, a large house and smaller buildings were erected near the shore. Olmstead "truck logged", first constructing wooden-plank roads, as no gravel was available and he had no machinery to crush rocks. Lumber for the roads was cut at the sawmill at Tom Lake. Some of the roads were necessarily very steep, and trestles were required to cross the numerous gullies. Olmstead managed to overcome every difficulty; a determined individual, he was known to be equally stubborn at sea, letting neither fog nor storm deter him from setting forth in his boat on a scheduled date.

Redonda Bay's school, with Margaret Sinclair as teacher, was held in the Olmsteads' old home until the Olmsteads built a schoolhouse in 1941. Community celebrations were held in the schoolhouse, including the wedding reception of one of the Olmstead sons.

During the peak of the canning and logging industries, the population of the bay rose to around 100 people, most of whom lived in houses built and owned by the canning company. Redonda Bay was a busy place during these years. Besides the Union Steamships there was a steady stream of vessels calling regularly: Coast Ferries vessels,[9] the Imperial Oil barge, the Royal Canadian Mounted Police boat, floating library, travelling dentist, various mission boats, pleasure craft and fishing boats. The *Sundown, Marshall Wells* and *Caribou* were like travelling stores, carrying samples of toys, gifts and household goods for families, as well as supplying logging equipment and other industrial needs. The Olmstead boys and their wives remember gathering in one of their homes to play canasta by the hour while waiting for the Union Steamship to arrive on its uncertain schedule.

The Olmsteads sold out their logging company in 1949 to Giroday Sawmills, a large company that had tractors and heavy machinery which allowed them to reach the rich timber valleys east of Jack Lake (now Baile Lake.) By the mid-1950's lumber production had risen 100%, with eight truck-loads of logs brought out in a day. During this boom a government wharf was built, and air travel was provided for emergencies, as well as for "movie and mail flights" to and from Vancouver. There was even a flurry of mining activity at this time when loggers accidentally struck molybdenum while putting through a road on the slopes of Mount Perritt. West Redonda was extensively prospected but the claims yielded little of value.

By 1956, Redonda Bay's second boom had petered out. Francis Millerd's cannery closed down, although it kept its store open until 1964. Boats were equipped with cold storage facilities by then and numerous canneries and cold storage plants were no longer needed; companies serving the area were centralizing around Vancouver. Giroday sold out around the same time to L & K Lumber, which wanted logs for its North Vancouver sawmill, but in 1966 it too was forced to close down. A forest fire had destroyed much of the mature timber, and the government refused to allow the second growth to be cut until it reached maturity.

With no industries to attract permanent settlers, the population of Redonda Bay fluctuated. People came and left again when they found no means of making a living. Squatters lived in the old buildings for a time, but these were considered fire hazards and eventually were razed and burned. A half-century of industry and settlement at Redonda Bay came to an end.

Today, Redonda Bay is the site of a minimum security camp operated by the British Columbia Forest Service.

Twenty selected inmates from provincial correctional centres throughout the province are employed on a work release basis. Their task is to thin out heavily timbered areas to encourage mature forest growth in the years to come.

CHAPTER 5

Stuart Island

"A round island, 3 or 4 leagues in circuit,
lying before the entrance into Bute's Canal,
received the name Stuart's Island."[1]

Stuart Island, rising abruptly to 880 feet almost at the
water's edge, is made historically famous by the sketch in
Captain Vancouver's journal which shows a point opposite the island in the mouth of Bute Inlet, called by his exploring party "The Village of the Friendly Indians." Once
an anchorage for Indian dugouts, Stuart Island now provides wharves and landings for the numerous yachts and
launches of sports fishermen who come to fish the Yaculta
Rapids that lie along the western shore of the island.

Vancouver authorized the name of the island in July
1792, in honour of John Stuart, third Earl of Bute
(1713-1792), a close friend of George III, and gave Bute
Inlet its name from the earl's title. Admiralty surveyors
around 1863 named the northern rapids after the island
of Arran in the Firth of Clyde, county of Bute, Scotland.
Galiano named the island *Cevallos*, and Bute Inlet he

named *Quintano* after one of Malaspina's officers. The Arran Rapids were called *Angostura de los Commandantes* because Galiano and Valdes examined them by longboat before attempting passage with their schooners. However, only the English names have been retained.

Johnstone, the master of the *Chatham*, was sent by Vancouver in the ship's longboat and cutter to explore Bute Inlet. The party had progressed a short distance up the inlet when two canoes containing about a dozen Indians came to barter bows and arrows for small trinkets. When the English boats came back down the inlet from Waddington Harbour on June 30, the Indians paddled out from their village once again. Vancouver repeats Johnstone's story: "Here was found an Indian village, situated on the face of a steep rock, containing about one hundred and fifty of the natives, some few of whom had visited our party in their way up the canal, and now many came off in the most civil and friendly manner, with a plentiful supply of fresh herrings and other fish, which they bartered in a fair and honest way for nails. These were of greater value amongst them, than any other articles our people had to offer."

This was the Village of the Friendly Indians, near Arran Rapids, the narrow channel between Stuart Island and the mainland. The boats then entered the Yaculta Rapids and whirlpools where the current was so swift that they were unable to row and were forced to haul the boats by a rope along the rocky shores of the passage. "In this fatiguing service," wrote Vancouver, "the Indians voluntarily lent their aid to the utmost of their power, and were rewarded for their cordial disinterested assistance, much to their satisfaction."[2]

After the departure of the English for Discovery Passage, the Spanish schooners started up Lewis Channel, tacking from shore to shore to avoid cross currents, making little progress, and finally taking to the oars before arriving at Stuart Island. "We anchored at nightfall between the coast and the island which we called Cevallos . . . from a settlement which was on the island, three canoes came out with as many natives in each one, and made their way to the *Sutil*, where they were treated with much friendliness."[3]

The Spaniards described the Indians as of medium height, well made, robust and of dark colour, and in features, language, dress and arms much like those of the interior of the strait. The Indians were probably Salish, as this was before the southern invasion of the Kwakiutl. Stuart Island later was Kwakiutl territory and the Salish had moved farther up Bute Inlet.

The Indians warned the Spaniards that evil men would murder them if they went up the channel and urged them to visit their own village instead, "showing so consistently friendly and compassionate a spirit and such disinterested affection that we were unable to thank them sufficiently."[4]

Galiano and Valdes continued up the east side of Stuart Island, anchoring first near Henrietta Point and calling their anchorage *Robredo* after an officer of Malaspina, and farther up naming their anchorage *Fondeadero de Murfi* after another of Malaspina's officers. From here the two commanders embarked in the longboat to examine the Arran Rapids before circling the island. They heard the roar of the waters from afar and were amazed by their swiftness when they came in sight of them, estimating their speed at 12 miles an hour. "The aspect is a most strange and picturesque one. The waters look like a race from a waterfall, and on them great numbers of fish are constantly jumping. Flocks of seagulls settle on the surface at the entrance of the channel and after allowing themselves to be carried to its end by its rapid course, fly back to their original position. This not only amused us, but also afforded us a means to gauge in some measure the velocity of the current."

The Indians advised them not to attempt the rapids in their longboat; their own canoes had been caught in the whirlpools and swamped when they tried to shoot the wild water. When they understood that the Spaniards were determined to tackle the rapids in their sailing vessels, they predicted disaster, but told them the current was less swift when the sun reached the top of a high mountain on the mainland. The Spaniards anchored near the northeast point of the island to wait for this turn in the tide and the natives brought them fresh salmon and sardines during the wait, warning them yet again of the dangers they faced. "For this humane and benevolent attitude," said the captains, "we continued to call them 'Good Indians' and we strove to give them whatever we knew might contribute to their satisfaction and comfort."

Once into the narrows, they shot down at an "extraordinary rate", steadied by the fore topsails as the wind blew against them in an opposite direction. They tried to avoid the left side, which would take them to the Yaculta Rapids where their officers Vernaci and Salamanca in an earlier exploratory trip by longboat had seen violent whirlpools in which the water sank more than an yard, but the current kept forcing them from their chosen path. As they shot into the Dent Rapids the *Sutil*, having tacked near an island, went about on the opposite tack and was caught in a strong whirlpool that turned her round three times with such force that "those on board were made giddy." Galiano wrote in his diary: "A scene never before witnessed by any of those present, it unavoidably caused great laughter, not only among those who were in danger, but among those who were momentarily expecting to be."[5] They succeeded in freeing themselves by rowing with full strength, and finally anchored at nightfall by a small island. They named the island where the *Sutil* had turned about *Vueltas* ("many turns".)

At night the wind increased, whistling in the tops of the trees, while the violent rapidity of the waters caused "a horrible rumbling." It was, they said, "a terrifying situation."

*   *   *

On the day that British Columbia joined Confederation, July 20, 1871, the Canadian Pacific Railway began its survey of the province to decide on a suitable terminus for the transcontinental railway. By 1872 the surveyors had reached Stuart Island, chosen as their headquarters while they explored the neighbouring islands and channels "to ascertain how far it may be practicable to reach Victoria, Esquimalt, and other ports on Vancouver Island by a continuous line of Railway from the mainland."[6]

To reach Vancouver Island, the plan was to lay a bridge across the Arran Rapids to Stuart Island, another from Stuart across Cordero Channel to Sonora (spoken of as Valdes, since Sonora, Quadra and Maurelle were all called Valdes at this time) and over Seymour Narrows via Maud Island to Vancouver Island.

The H.M. gunboat *Boxer* and H.M.S. *Scout* explored Bute Inlet, and later when the *Boxer* returned and was anchored July 11 off Stuart Island, Marcus Smith recorded: "I immediately sent a messenger to a village of Eucletah Indians, a few miles up the strait asking the chief to bring down a number of his people who had applied to us for work and whom we now wished to engage . . . we engaged the chief and twenty of his tribe to go with their canoes to the head of Bute Inlet, and thence up the Homathco river with supplies for the surveying parties."[7]

This was 1872. Provincial Museum maps put the Kwakiutl on Stuart Island and around the mouth of Bute Inlet after 1850. (Kwakiutl refers to a number of tribes who speak somewhat similar dialects of the Kwakiutl language. Originally, the term *Kwakiutl*, or *Qagyuhl*, referred to the Fort Rupert tribes. *Kwawkewlth* is the Indian Agency's term for the Southern Kwakiutl and Comox-Salish district. Some Kwakiutl Indians prefer the spelling *Kwawkgewlth*, and many refer to themselves by tribal terms such as *Nimpkish*. "Why do they call us Kwakiutl?" Louise Hovell of Cape Mudge once complained. "We're Lekwiltok!" *Eucletah* or *Euclataw* and *Yaculta* were variations used by whites for the Lekwiltok band of Kwakiutl-speaking Indians but often were used, inaccurately, to mean any Southern Kwakiutl tribe.)

Marcus Smith wrote: "These Eucletahs are a warlike tribe, and, holding the narrow channels about Valdes and Thurlow islands, were formerly a terror to the other Indians and early settlers on Vancouver Island. They are finely built, strong and active, and they seemed anxious to work for us, those whom we did not engage exhibiting great grief at being left behind."[8]

The Victoria *Colonist* of August 10, 1872, printed a news item "From Mr. Michaud's Party" (the surveyors) saying they had conversed with a Captain Collins of the sloop *Duncan* who told them an easy crossing could be made at a point where the depth of water was only 52 feet. "The whole party are confident that the Narrows may be bridged."

Unfortunately, the survey plans were lost through fire, so in 1875 Marcus Smith made a second survey, travelling by canoe with an Indian crew. "This work was done in a canoe in the worst season for navigation, when as we afterwards learned, the Pacific coast was strewn with wrecks," he reported. Smith came to the conclusion that a steamboat could take a railway train on board and pass from Bute Inlet through the cross channels dividing the "Valdes Islands", to Vancouver Island. However,

Engineer-in-Chief Sandford Fleming decided both ideas were impractical and plans to make the crossing in this area were abandoned.

Anderson Secord was settled in the district in 1898, the Voters List placing him as a rancher at Amor Point in Bute Inlet. The 1907 Voters List gives Stuart Island as his residence, and his occupation as fisherman. He is still listed as a fisherman on Stuart in 1927.

The directory for 1899-1900 describes Stuart Island as "a post settlement on Stuart Island at the mouth of Bute Inlet, 100 miles up the coast from Vancouver," and lists one person: D. Chapman. August Schnarr, logger and cougar hunter, says there were only three people living on Stuart Island when he went there in 1909. The 1910 directory shows a logging camp established on Stuart Island and reached by Union Steamships. Ickes Henry Asman, farmer, is listed. A possible relative, Frank Asman, with occupations "fishing, ranching and logging" is in the 1919 directory, having pre-empted lot 1583 by the present Asman Point.

Settlement was mainly on the west shore around Big Bay, a good area for fishing the rapids. Charles H. Smith appears in the 1914 directory as fisherman, along with Albert Smith, farmer, who pre-empted lot 1579, between Kellsey and Whirlpool Points in 1913.

MacKenzie's pre-emption in 1911 was on the site of today's Stuart Island post office and settlement, but he cancelled this in 1918. George Bruce took it over, abandoned it in 1923 and finally bought the 76.4 acres later in the year. W.V. Willcock in an article in the *Colonist* of June 28, 1953 writes: "The landing has been used by the Indians for ages, but it was about 50 years ago that any attempt was made to settle down in this district. A half-breed was the first settler. Then, after his death, a man by the name of Bruce took it over and for years it was known as Bruce's Landing." A post office opened on February 1, 1919, with George Bruce as postmaster, and the directory for that year gives the "landing name" of the island as Bruce's Landing. Bruce ran the general store and was also a fur buyer and fish dealer. Henry Ives had a little store as well, built on a float and moored to the wharf. If a fish run occurred some distance away, Henry hired a gasboat to tow float and store to the scene of the run, to serve the fishermen.

Another early settler was Kellsey R. Moore, fisherman and rancher, who pre-empted lot 1578 in 1913. Kellsey Point is named after him. Mount Muehle, 1710 feet high, near the central part of Stuart Island, is named after William Muehle, another fisherman and rancher.

Fisherman Henry Maurin, a French-Canadian married to an Englishwoman from Liverpool, pre-empted in 1913 a half-mile from Bruce's Landing and built a large house that was used as a Red Cross outpost during the 1940's. Besides acting as guide to the American author, Stewart Edward White, who often came to fish the Yacultas in his power launch *Dawn*, Henry fished for the huge ling cod that frequented the rapids. He delivered them alive in his fishbox to the logging camp, killing and cleaning them on the wharf and trundling them by wheelbarrow to the waiting logging train. Big devilfish or octopi lived under the wharf and devoured the offal as he tossed it into the water. A *Province* news article of September 27, 1925, says that one of these huge creatures, nine feet across its tentacles, was caught by Vancouver sports fisherman Henry Hoffar, from under a log boom in the Yacultas. (He also reported a 60-pound salmon and a 60-pound cod as part of his catch.)

"Cabin fever" was common among dwellers in a solitude they found themselves unfitted to endure. Henry Maurin was called on to deal with a deranged man who believed a voice had told him to take seven wives. The man had entered the home of a married woman and asked her to be the first of the seven. She panicked and called her husband who tried to oust the intruder, but the demented suitor seized an axe and wounded both the husband and the husband's mother. Neighbours called Henry, who arrived carrying a long pike pole. He offered to drop the pole if the man would discard the axe, and when both had disarmed themselves he conveyed all of them to the hospital, the injured pair aft and the wife-seeker closeted with Henry in the wheelhouse. Many men grew desperate for wives in those lonely areas. One Frenchman kept a sign "WIFE WANTED" on a tree by his cabin, but had no takers.

Fishing and logging were the local industries in 1919, with the population estimated then at 100. The South-gate Logging Camp appeared in the 1919 directory, with T. Bernard as manager, and there was also a dogfish plant run by J.C. Jardine. No mining activity had been reported on Stuart and none was anticipated by Mining Department maps. A black miner sank a deep shaft in Big Bay (formerly Asman Bay) with pickaxe and shovel but found no minerals. The shaft is still there.

The first school, with an enrolment of two boys and seven girls, was established late in 1927, with teacher Miss E.M. Brown receiving an annual salary of $1020. There was a big logging camp on the island, with 10 or 12 families with children. Alan Greene called at Stuart Island in the *Rendezvous* in 1932. He found many of the fishermen suffering from hard times, with codfish prices down to 1½ to 2^e a pound, and some of those in need had been given work on a trail to the Asman Bay school run by Miss Hartshore. Children attending the school were taken from their homes by rowboat and gasboat to the start of the trail, the boats leaving early in the morning to catch the slack tide through the Yaculta Rapids. All events had to be timed by the tides.

Mrs. Joseph Dick, formerly of Heriot Bay on Quadra,

was w :king that year as assistant to store manager M.V. Willcock of the Stuart Island Trading Company. Willcock in a 1953 newspaper article said there never was a building boom on Stuart Island, and estimated the population in that year as still around 100. He told of the little store that carried everything from bed pans to lumber for coffins, this last requested by the Indian reserve at Church House. Sometimes, he said, the Indians paid for their goods with a seal's nose, for which a five-dollar bounty was given by the government.[9]

The changeover from commercial to sports fishing on Stuart began with Bert and Mary Brimacombe, who came to the island in 1935 and stayed in Big Bay for over 40 years. They lived first in a little shack on the beach and were part of the lusty era of those days. There was a logging camp in the bay and the men were all loggers or fishermen.

Brimacombe tells a story of an old logger, Fred Lobisher, one of the best chutebuilders on the islands, who was called in often to supervise chutebuilding long after he was too old to do the actual work. He was a huge man, weighing almost 300 pounds, fond of drinking and dancing. One night, a dance was held in the community hall, a building that had been towed down from some camp and anchored on the opposite side of the bay. Halfway through the evening, Lobisher found himself giddy from overindulgence in his two addictions and asked Brimacombe to help him home, as he needed to negotiate a narrow footbridge over a stream to get to his shack. Brimacombe persuaded another neighbour to join him on the walk as one man alone could never lift Lobisher if he fell down during the trip. Supporting him, one on each side, they managed to get him home safely, undressed him with great difficulty and tucked him into bed.

It had been an exhausting experience, and the two men stopped off briefly at Brimacombe's for a drink to revive themselves after their efforts. When they got back to the dance, the first sight that met their eyes was Lobisher, fully dressed and preparing to start his revels all over again! After the dance, Brimacombe and his friend walked home together. They came to the footbridge, and there beneath it in the shallow water lay Lobisher, groaning pathetically. "Ah, let him lie," said the exasperated neighbour, but eventually Lobisher's pitiful cries softened them and they relented, hauled him up, and got him home and into bed once more.

In 1952 a plane landed in Big Bay and disgorged an American tourist who had heard of the big fish to be caught in the Yacultas. There was no hotel, so the Brimacombes made up a bed for him on their couch. The next season more Americans arrived, and the Brimacombes managed to squeeze them into nooks and corners. Commercial fishing was becoming unprofitable but there were still enough fish to supply sportsmen, so the Brimacombes spent the winter building a cottage, attached to the new home they had built earlier but completely self-contained, with two baths and room to put up ten tourists. They did no advertising, but the cottage was full every season and their guest book included such names as Roy Rogers, Ethel and Robert Kennedy, Governor Albert Rosellini and Senator Jackson.

Boats were soon packed four and five deep around the floats in July and August, and even the anchorages were crowded with tourist craft. Gulf Airlines services the island twice daily now and Kenmore Charter planes from Seattle arrive with passengers three or four times a day during the summer months. Where formerly 90% of the boats in the bay were commercial fishermen and 10% tourists, now the reverse is the rule. Fish buyers, fish camps and ice packers are all gone; the tourist trade has taken over. Most local residents have moved out, and the school is expected to close shortly.

Bert and Mary Brimacombe retired to Vancouver Island in 1978 but their son Terry and his wife Louise are running the resort. There are four resorts now on Stuart Island, at the foot and head of the Yacultas. With charts and tide books to guide them, today's boatmen suffer few of the terrors of those first explorers who ventured so dauntlessly into unknown waters.

CHAPTER 6

Maurelle Island

Maurelle Island, about 20 square miles in area, is divided from Read Island by narrow Hoskyn Channel. The channel contracts at one point to a shallow, rock-choked passage marked five feet minimum on hydrographic charts but becoming almost dry at low water. Most of Maurelle is rocky and broken but there are areas of arable land in the south where there have been surveys and settlement.

Known for over a hundred years as Middle Valdes Island, Maurelle in 1903 received its present name honouring Francisco Antonio Maurelle, a Spanish naval officer who accompanied Quadra in the *Sonora* on his exploratory voyage up the northwest coast in 1775. They travelled far enough north to see Mount Edgecumbe in Alaska, and a short account of the voyage was published in the government gazette in Madrid and copied in London newspapers. Another expedition was ordered for 1779, with Quadra and Maurelle together again, in the

Favorita. This time they reached Prince William Sound, and a few years after their return Maurelle published his *Compendio*, describing the Spanish voyages up to 1790. The section about the northwest coast voyages was translated into English, giving Maurelle an international reputation. The Spanish viceroy wrote him a flattering letter, with detailed, confidential instructions for a voyage to the California coast, Juan de Fuca Strait and Alaska.

Unfortunately, Maurelle fell ill. Jacinto Caamano was slated to take his place but, while riding down a steep slope on the 70-mile journey to the naval depot at San Blas to embark, his horse stumbled and Caamano fell and injured himself. He too was unable to undertake the expedition. Malaspina, the famous Italian explorer in the service of Spain, suggested that he could lend his officers Galiano and Valdes with the schooners *Sutil* and *Mexicana*, to substitute for Caamano in a part of the original commission. This started the two explorers on their famous voyage in 1792 to the northwest coast.

When they reached the vicinity of Maurelle Island, the Spaniards anchored at the *Islas Tres Marias* (now Rendezvous Islands) off Maurelle, to rest after struggling against the currents of Lewis Channel. When the tide was right for them to proceed, the wind died down, and they were forced to furl their sails and row all the way past Maurelle to Stuart Island. "The tide gave us little help," they said, "and at times some of the eddies were so strong against us that with the oars we stemmed them with difficulty."[1]

* * *

Tom Bell of Quadra wrote in 1895 that bears were reported on Middle Valdes Island (Maurelle), sometimes crossing to Read Island at Hoskyn Channel, "a strip which dries at low water." At this time, he said, there were no settlers on Maurelle or Sonora.[2]

A 1912 survey map shows only six surveyed lots on

Maurelle. The Brunette Sawmill Company, managed by
H.L. De Beck, received a Crown Grant for lot 174 in the
centre of the island in 1894. Frank Fox took the lot above
174 in 1910, and the lot above that went to A.L. Clark.
Alex Russell, the logger who bought up land on several of
the islands, pre-empted lot 18 in the south near Antonio
Point in 1883, and there was a timber lease above and
east of this. Alan Hall pre-empted a small lot by the
narrow part of Hoskyn Channel in 1910. Just south of this
is an Indian reserve.

The south end of Maurelle was considered best for agriculture, most of the remainder being rocky and broken.
By 1916 much of the island had been logged off, leaving
many stumps, while fires had cleared away the
underbrush, so it was hoped that further clearing would
be fairly easy for settlers. A squatter, F. Kilpatrick, had a
good house and garden in the southeast at this time, while
another squatter, T. Keefer, next to him, had a home and
garden and also owned 16 head of cattle. The nearest
store was at Surge Narrows on Read Island.

John Armstrong with his wife, young son and
daughter, arriving on Maurelle from Berkeley, California
in 1928, found their new environment a complete contrast to life in a bustling city. "Main Street" on Maurelle
was a lonely trail that wound through the forest, just wide
enough for a wagon and team. There was no traffic problem, for there was only one wagon and team on the
island. To get to other islands, residents rowed, or asked
for the services of one of the two Read Islanders who
owned small gasboats. Gasboats were not common
among the upper islands until the late 1930's.

Cougars were numerous on Maurelle. If settlers were
wise, they took small animals inside before dark as
cougars would creep up and snatch dogs or cats from the
doorsteps. Mrs. Armstrong looked out of her window one
evening and saw a cougar sitting on a stump only a few

feet away, while a neighbour, hearing snarls and squeals in the night, opened his door to find a cougar about to pounce on his dog. He shot the big cat, but was too late to save his dog. With the predators about, there were few deer on the island, and venison was a rare treat.

Logging and fishing were the main occupations during the 1930's. Six mining claims were filed on the island in 1920 but none proved productive. Bancroft, the geologist, felt that some of the post-glacial marine clays on the south side of Maurelle, about one and a half miles from Surge Narrows, would probably be suitable for the manufacture of bricks and tiles.

The Patterson family, with a farm about two and a half miles inland, shipped cream from their cows via Union Steamship to Vancouver for several years. This entailed hauling the cans of cream to the beach in the island's one wagon, and then ferrying the cans by rowboat to the government wharf at Surge Narrows on Read Island, where the steamship called to load and unload mail and freight.

A school opened in 1929, with Elizabeth Mrus coming from Ladysmith to teach. It was arranged for her to board with the Armstrongs, but she had been there only a few days when the house burned to the ground and all her possessions and those of the Armstrongs were destroyed. Replacing a house and furnishings on a wilderness island in those days was a frustrating task. Not only was money scarce, but almost everything had to be replaced through mail order houses, with long waits for delivery. John Armstrong built a new home with the help of neighbours, who also boarded the teacher until the Armstrong house was readied. Despite her unhappy experience, Elizabeth Mrus stayed with the Armstrongs as teacher on Maurelle until the school closed four years later.

The first schoolhouse was a small cabin owned by a

bachelor settler, Bill Heinbokel. A few months later, school was held in a large room in another building in which Heinbokel was living. A year after the opening of the first school, an official schoolhouse was finally completed, with enough children mustered from the Armstrong, Patterson, Nelson and Morris families to complete the quota. A great blow was the settling of the Oliver Jones family on a site that was too far from the school for their children to attend, but the McIntyres swelled attendance for a time when they moved from Waiatt Bay, which had no school, and came to live in Heinbokel's cabin while Maurelle's school operated. Every school day, Sadie Dennis rowed over with her little girl, Violet, from a small island nearby. The Keefers, with a baby girl, were looked on as future prospects. By June 1933, however, attendance at the little Maurelle school had dropped below the required quota. In September, there were still too few children for the school to open, and it remained closed for good.

These were Depression years, hard years for a woman coming from the city to the wilds, and young Mrs. Armstrong was grateful for advice from Sadie Dennis, who had lived all her life on the islands and helped her to cope with the problems of primitive living. Neighbours were few but friendly, island living lacked luxuries but was healthful and inexpensive, and the atmosphere was one of freedom and unrestraint. Former residents remember their life on Maurelle as arduous, but call it an experience they would not have wished to miss.

For the Armstrong children, it was an idyllic time of simple pleasures, whittling boats from driftwood and sailing them down the fast-running creeks, or building dams to create miniature harbours. All about their clearing grew giant bracken, glossy salal and huckleberry bushes, and above these the fir and cedar trees mingled their sharp, bracing smell with the salt sea breeze from

Hoskyn Inlet. As teenagers, the children were kept busy helping with the care of the farm animals and with the canning of home-grown fruit and vegetables on the wood-burning stove. Meat from the farm livestock was also canned, and was supplemented by a year-round supply of fish from the sea.

While the school was in operation, an occasional dance was held in the building, with a group from Heriot Bay on Quadra providing music, and boat-loads of people coming from other islands, the women bearing food. Small children were put to sleep on benches that lined the walls. Ethel Armstrong Redford remembers how the lively polka music still sounded in her ears when, as a 12-year-old girl, she sat learning her lessons in the quiet schoolroom on the days following the community dance.

The little Heriot Bay band was composed of Ture Krooks, accordion, Lars Krook, violin, Eric Krook, zither, and Bill Law, drums. The four played for dances on many of the islands in the late 1920's and early 1930's, travelling in a boat with a one-cylinder, five-h.p. engine that required them to leave the bay around four in the afternoon to reach places like Stuart Island by eight. They played until dawn for the tireless islanders, returning home richer by 50¢ apiece.

In the 1930's, logging companies such as Niemi were active on the island again. The Patterson and Keefer farms are deserted now, and the Armstrongs have moved off the island. About ten people live on Maurelle, just above Surge Narrows. The few children on the island attend the Surge Narrows school in the big Quonset hut on Read.

Just north of Antonio Point on the southern tip of Maurelle there is a provincial recreational reserve. Here there are two small anchorages where boats can wait for a favourable tide before entering the waters of Surge Narrows.

CHAPTER 7

Sonora Island

Rugged Sonora, 56 square miles in area, rises from Cordero Channel to altitudes of over 3,000 feet. There are sheltered bays in the north and south, but three sides are surrounded by rapids: Hole in the Wall, Okisollo and the Yacultas. In the narrow channels that separate it from Stuart, Maurelle and Quadra Islands, the current reaches a speed of ten miles an hour, often with an overfall of two to four feet. Ships in the early days tried to avoid the rapids that threatened to crash them against the sheer, high cliffs of Sonora, but sportsmen today, familiar with tides and currents, find the Yacultas an irresistible attraction with their promise of salmon weighing up to 90 pounds.

After Valdes Island was found to be three islands instead of one, these were named Sonora, Maurelle and Quadra in 1903. The combination of names was appropriate, for Quadra and Maurelle had sailed up the northwest coast in the *Sonora* in 1775. As they passed through the Gulf Islands, seven of the *Sonora*'s crew went ashore for water on July 14 and were attacked by natives. All seven were killed. The Spaniards named a small island near the site of the tragedy *Isla de Dolores*, "Isle of Sorrows." The *Sonora* continued northward, sighting the mountain that Cook later named Mount Edgecumbe. The Spaniards named it *San Jacinto*, as they first saw it on St. Hyacinth's name day.

The Yaculta Rapids on Sonora's east coast almost defeated both English and Spanish explorers in 1792. Johnstone and Swaine attempted them in their cutter and launch and said: " . . . the Water rushed in Whirlpools with such rapidity that it was found extremely difficult even to track the Boats along shore against it, & this could hardly be accomplishd had it not been for the friendly activity of the Natives who in the most voluntary manner afforded them every assistance in their power, till both Boats were safely through these narrows . . . "[1]

Galiano and Valdes, having got through the Arran Rapids in their schooners, had tried to keep from being drawn into the Yaculta and Dent Rapids but the current kept carrying them away from the right-hand shore and into mid-channel. The *Mexicana* got through, but the *Sutil*, swept away by an eddy, almost grazed the rocks that jutted out from the shore and was swept into the *Canal de Carvajal* (Cordero Channel) between Sonora Island and the mainland where she turned around three times. A cable was run out to the east point by the boat, but the schooner was caught in another whirlpool and turned about again, tearing the cable from the hands of those trying to make her fast. The Spaniards found the cross currents and counter currents baffling; the *Sutil* tried four times and was carried back each time.

Eventually, having become familiar with the passage, the *Sutil* made it through by rowing strongly, and the vessels continued up past the east side of Sonora, hugging the mainland, and into Nodales Channel west of Sonora, anchoring a cable's length from land. Here Galiano stayed while Salamanca explored Loughborough Inlet (*Brazo de Salamanca*), after which they sailed north and around Thurlow Island and out to Johnstone Strait.

Captain Pender, in the hired surveying vessel *Beaver* in 1863, named Cameleon Harbour on Sonora after H.M. screw sloop *Cameleon*, stationed at Esquimalt. Edward Point and Hardinge Island at the entrance to the harbour are named after the *Cameleon's* commander, Edward Hardinge. The *Cameleon* joined the *Forward* in an expedition against the Lilmalchey, a Salish tribe, after the Indians murdered Frederick Marks and his daughter on Saturna Island. Various names in Cameleon Harbour honour the *Cameleon's* officers on that expedition: Bruce Point, Binnington Bay, Tully Island, Handfield Bay and Greetham Point. Maycock Rock and Piddell Bay are named after two young officers of the flagship *Sutlej*, stationed in Esquimalt.

The Canadian Pacific Railway survey of 1872 considered a route from Stuart Island across Sonora, Quadra and Maud Islands to Vancouver Island. It is interesting to speculate upon the difference in development that such a railway line might have made to Sonora Island, but the idea was eventually abandoned by the government on the advice of Sandford Fleming. Apart from the difficulty of bridging channels, Marcus Smith, who wrote the Report of Progress said the line would necessarily follow "rather a circuitous course, to avoid high rocky hills."[2]

The heavily-timbered slopes of Sonora attracted logging companies as far back as 1888, when Leamy and Kyle obtained a 21-year timber lease at the head of Cameleon Harbour. Pacific Coast Lumber Company and Chemainus Lumber and Manufacturing Company had leases in the 1890's near Owen Bay and on Venture Point on Okisollo Channel. In the early 1900's, the Vancouver Lumber Company and Canadian Puget Sound Lumber and Timber Company obtained leases in Cameleon Harbour and around Thurston Bay up to Florence Lake.

John Antle in 1906 visited the Okisollo camps, these being regular stops for the Columbia Coast Mission boat *Columbia* which had only been in operation a year at that time. Antle arrived at Tomlinson's camp on a snowy day in January and crossed with some difficulty a boom of logs covered in five inches of snow to reach the shore and the logging camp. The camp manager, J. Graham, showed him around the neatly-painted bunkhouse, cookhouse and office, and also pointed out the livestock: cows to supply fresh milk for the crew; oxen, pigs and chickens. Five Russians from Manchuria were working in the camp.

There was a government forestry station at Thurston Bay on Sonora, and a forestry lookout on top of Mount Tucker. Before the beginning of World War One a telephone line connected the forestry station with Shoal Bay on East Thurlow and with Rock Bay on Vancouver Island, the Columbia Coast Mission hospital site. A survey report in 1916 to the Minister of Lands said that Sonora had been swept by fire a few years previously and the ground was covered with wild blackberries, the usual aftermath of forest fires. An old logging skid road ran along Johnstone Strait on Sonora's west side.

Single men came first to live on Sonora, settling at Owen Bay and working mainly at hand logging. Hand logger licences were issued to Indians or to men eligible for the Voters List, and were for one year only. The area had to be inspected and approved by the Forest Service, and only muscular power could be used to fall the trees; no machinery could be employed. When first issued in 1886, licence fees were $10; this was raised to $25 in 1908. Very few hand logger licences are asked for today.

Men who spent their lives logging often felt lost when old age ended their working days, and they clung to areas where they had once been active. Hiram Corn dragged his house up a steep bank on Cordero Channel with his donkey engine, and surrounded the house nostalgically with his old logging gear. It was a lonely and worrisome life for his wife. She seldom had visitors, and was torn by anxiety when old Hiram, weak and sickly, set forth in his rowboat every two weeks for the long trip to Shoal Bay on Thurlow Island to pick up groceries. Pete McDonald ran the store and post office then, in 1918 and into the 1920's. The trip took Hiram hours of rowing and he would arrive home exhausted. His wife was always afraid that he might collapse and be swept into the Yaculta Rapids, but he refused to leave the familiar surroundings where he felt at home.

One night he suffered a stroke. His wife signalled frantically to passing boats for help, but it was some time before one came to her aid. Her troubles were not over, for, after leaving her husband in the hospital, she was shipwrecked on her return trip and spent the night on a small rocky island. When her husband died, she sold her property and moved from Sonora without a backward look at the home where she had been so lonely and distressed for ten long years.[3]

Fish and game were free for the taking, but men worked at logging, fishing and trapping to bring in the cash needed to cover other living costs. August Schnarr, who added cougar hunting for bounty to these pursuits, was born in Centralia, Washington, in 1886. He came to Gastown when he was 21 and in 1909 travelled with a friend on the *Cassiar* to Cracroft Island to work in a small logging camp. The gnats were ferocious in the woods, attacking in huge clouds, getting into eyes and ears and raising great welts out of all proportion to the insects' minute size. They were too much for Schnarr's friend, who returned to Gastown, but August stayed on, owning one of the Rendezvous Islands for a time, then logging and hunting up Knight and Bute Inlets and settling in Owen Bay on Sonora with his family in the early 1930's.

Schnarr's tame cougars on Sonora were famous. He had killed a female cougar and noticed that it had been nursing recently, so he hunted about until he found the cubs and brought them home as pets for his three daughters. The cubs were kept in a pen during the day and brought into the house at night. The children romped with them, and carried them about in their arms even after they had grown to almost the same size as the girls themselves. The cougars would purr and run to be petted when they saw the girls, but uttered angry hissing sounds when a stranger drew near. When they reached maturity they were kept on light chains, long enough to allow them to play together. Finally they grew strong enough to break the chains and would run off to molest chickens and other livestock. They always came back out of the bush, however, if the girls called to them. Heavier chains were finally used to protect neighbouring livestock. Despite their unnatural existence, one cougar lived 3½ years and "Girlie" lasted 6 years.

Now 91, still sturdy and independent, Schnarr lives alone at Heriot Bay. The years that he spent travelling in dugout, rowboat and gasboat through the rough waters of Bute Inlet, the Yaculta Rapids and the rapids outside Owen Bay have been turned to good account in his hobby of boatbuilding. He designs all the boats himself, to best combat the vagaries of the northern rapids.

Owen Bay cuts deeply into south Sonora, off Okisollo Channel. The wild, rocky central portion of the island was frequented by cougars, and no roads connected the south and west bays. A trail (overgrown now) that ran from Owen Bay to Cameleon Harbour was used mainly by cougar hunters. As a result, Owen and Thurston Bay settlers thought of themselves as separate settlements rather than parts of a single island. Several families came to settle around Thurston Bay. A school for their children was listed for 1928, with Miss G. Galliford as teacher. The listing continued until the mid-1930's.

The Bentleys were the first family to settle in Owen Bay, and they opened a small store there. Shortly afterwards, in 1925, Logan Schibler and his Norwegian wife, Gunnhilde, with their three children arrived from the United States, travelling in a 20-foot rowboat that Schibler had built from hand-split, cedar planks. Before settling at Owen Bay he had logged on mountainous little Raza Island, but the bay's rich, black soil and the abundance of fish and game encouraged settlement with its promise of a year-round supply of food. They were fortunate to fall heir almost at once to property belonging to an old Norwegian settler known as "Whiskey Harry" who offered to turn over his house and land to Schibler's Norwegian wife if she would care for him in his old age. Whiskey Harry's house became the first little school in Owen Bay in 1926 with Miss Edith Procter as the teacher. Later, in the 1930's the school was a fish buyer's building on a float, towed down from Stuart Island. Alan Greene speaks of this "floating school" in 1937 when Mrs. Van der Est was the teacher.

The Schiblers were born pioneers, hardworking and versatile. In record time their land was producing an abundant crop, and Gunnhilde was putting up over 500 quarts of vegetables, fruit, fish, grouse and venison each year. Nothing was wasted. Duck feathers were saved to fill flour sack pillow cases that were lightly waxed with a quill to prevent the feathers from working through the sacks. Logan built and ran a mill on Mill Point and cut the lumber for a new little frame school in the 1940's. On his float stood a big wooden tub in which the fishermen soaked their linen fishnets to kill the bacteria that rotted them if they were left untreated. Today, nylon nets eliminate the need for these tanks. Schibler's power plant supplied electricity for the school and most of the residents and for a large house that he moved up to Owen Bay by float. The house became a focal point for community gatherings and dances as the bay built up, and often Logan and Gunnhilde collected strangers waiting for a favourable tide in boats tied up at the float and brought them in for a hearty meal at the Schibler table.

There was no official post office there in the 1920's and 1930's, but when the Union Steamship boats were running, "open bag" or "string bag" (unsealed) mail was dropped off at Waiatt Bay for the surroundings areas. In those days, the Union Steamship Company was permitted to use the Cape Lazo weather station's radio wavelength, and at stated intervals the captains would broadcast the position of their ships and the expected times of arrival at various ports. Schibler would listen to the broadcast on his boat radio, then run his boat over to Waiatt Bay to collect the mail, and sort and hand it out in the living room of his house.

If the weather was too stormy, the ship cancelled the Waiatt Bay stop. After the Union boats had several accidents travelling up the dangerous passage, an alternate route was chosen, with a stop at Chonat Bay, at the float of whichever logging camp was working on one side or another of the Okisollo Channel. Still later, in the 1940's, mail was put off at Surge Narrows on Read Island, and postmaster Clements went down to collect it, travelling with the tide, and making the return trip in the dark, as the boat arrival was always at night and freight and mail had to be picked up immediately. After the Union Steamship Company ceased its service, Gulf Lines ran the route a few times, as did a freighter, but then all service ceased.

In the 1920's, every little bight around the large bay

and along Okisollo had a small shack or two drawn up on the foreshore, and near Hyacinth Lake there were farmers raising vegetables, and cutting hay on the nearby swamp meadows for their livestock. One farmer paddled a raft across the lake and hiked to Owen Bay for his groceries. (Sonora gives Hyacinth Lake the English spelling, whereas Quadra holds to the French spelling for Hyacinthe Bay.)

Sonora Gold Mines Ltd. did some work in the 1930's in the northwest part of Sonora, but the yield was not considered important. Mining maps suggest that an area around Sonora Point, near Mount Tucker, might be promising.

Logan Schibler's daughter Helen married C. Gus Clements, who was one of the last of the homesteaders in the area. Clements ran a co-operative store on the waterfront at Owen Bay and a fish scow that stored fish brought in by fishermen until the packers called to collect the catch.

He also raised hundreds of mink, encountering peculiar problems due to the farm's seaside location. Occasionally a mink would escape and make for the water. The Clements would pursue it, Gus steering the boat and Helen using a dipnet to urge the runaway back to shore. The mink were raised in individual cages scarcely larger than the animal itself, but mink kits that had lost their mothers were more fortunate. They were nursed through infancy in a bed behind the kitchen stove, kept warm by a hot water bottle and a friendly mother cat. These house-grown mink were unable to adapt to cages and had to be kept in the house until old enough to be killed for their pelts. Like half-tame pets, they ran freely about the rooms, jumping on the kitchen table and sampling dishes that appealed to them.

The Clements left Owen Bay in 1951. That year, Logan Schibler's son Jack married a young schoolteacher,

Ella Steele, who had come to the bay to teach the ten pupils for a year. One year later, attendance dropped and the school was closed. Young Ella found it difficult at first to adjust to a community where there were no roads and where the time of paying a visit by sea must always coincide with the time of slack tide. Fortunately, the Schibler house was a lively Mecca.

When Logan died in 1957, Jack bought the property his mother had inherited from old Whiskey Harry; he ran the mill and fished and logged for ten years before moving to Vancouver Island. By then, it was apparent that the era of farming and settlement at Owen Bay was drawing to a close. Fast motors were available on the market, and boats could travel swiftly, bucking tides that defeated the little two-cycle engines of the early days. There was no need to live in isolated areas to be near the fishing grounds, and residents began to move out to populated islands where living conditions were easier. Today it is mainly transients who come and go in Owen Bay. A real estate company has bought up most of the lots and is holding them for speculation. A few have been taken by summer residents.

The Forestry station and other buildings are gone now from Thurston Bay, and farms are abandoned. The provincial government established a Class A marine park of 875 acres there in 1970. Some logging is still carried on in the vicinity, and several logging camps are situated in the south on Okisollo. Recently, a group of young people attempted to establish a farming colony on 150 acres on Sonora.

CHAPTER 8

The Thurlow Islands

For many years the two Thurlow Islands were busy logging and mining areas, the stamping grounds of mining men and grizzled prospectors, of loggers who worked and played with a reckless vigour, and of independent men who fished in their own boats and kowtowed to no master. Hotels at Shoal Bay were crowded; there was a cannery at Blind Channel. The busy days of industry are over, and the Thurlows, like Stuart Island, now host the sports fishermen who wait in Shoal Bay for favourable tides before tackling the Greene Point Rapids or the Yacultas.

Lower or East Thurlow Island is 43 square miles in area, its terrain mountainous, with a steep rocky shoreline. Upper or West Thurlow, separated from lower Thurlow by Mayne Passage, is 33 square miles and also mainly hilly, rough and timbered, with a few small lakes. As was the case with many of the islands, the Thurlows were originally named and charted as one land mass. Captain Vancouver, with his two vessels,

Discovery and *Chatham*, sailed up Discovery Passage *en route* to Nootka in July of 1792. On July 16, Vancouver wrote in his journal: "After we had proceeded about ten miles from Port Chatham, the tide made so powerfully against us as obliged us at breakfast time to become again stationary in a bay [Knox Bay] on the northern shore in 32 fathoms water. The land under which we anchored was a narrow island which I distinguished by the name of Thurlow Island."[1]

The name honoured Edward, Lord Thurlow, son of a Suffolk clergyman who began his career in a solicitor's office, was later appointed Lord Chancellor and was raised to the peerage in 1778. When the Duke of Grafton criticized the appointment of a "plebian" to the House of Lords, Thurlow made a proud and indignant speech in retaliation which has been called "a gem of English oratory." Lord Thurlow was tall and majestic in appearance, and it was said of him: "No man ever was so wise as Thurlow looks." When Thurlow Island was found to be two islands, the new names became East and West Thurlow.

James Johnstone, R.N., master of the *Chatham*, passed the Thurlows on the north early in July as he searched for a passage leading to the sea and the western shore of Vancouver Island. His party took time to explore Loughborough Inlet (which Vancouver later named after Alexander Wedderburn, first Lord Loughborough, Lord High Chancellor of England and an intimate friend of the Earl of Bute). Here, spending a night on a small island, they were wakened from their slumber by the flooding tide, which came in so suddenly that all were soaked with spray and too uncomfortable to return to sleep. On the way back, having reached the sea and confirmed the route as desirable, Johnstone, in the *Chatham's* cutter, and accompanied by Mr. Swaine in the launch, once more passed the Thurlows, this time on the south.

When Vancouver reached the Thurlows, he was

pleased with the bay on West Thurlow, which afforded
him good anchorage and an easily-obtained supply of
wood and water. Once again, however, he had no luck at
fishing. As at Desolation Sound, neither seine nor hooks
and lines caught them fish of any kind, and to add to
their mortification, when they passed an Indian village
the Indians tossed "ready-cooked" and fresh salmon to
them. "They seemed to have great pleasure in throwing
them on board as we passed their canoes," wrote Vancouver.

Squalls struck overnight. They double-reefed their topsails and plied to windward, "with little prospect of
making much progress, until we had passed Thurlow's and
Hardwicke's islands."

The Spanish explorers, Galiano and Valdes, having
taken a different route, circled Stuart Island in the *Sutil*
and *Mexicana* and came again into Cordero Channel
(named after Josef Cordero, draughtsman of Galiano's
expedition). Friendly Indians paddled alongside the *Sutil*
in two canoes and pointed out the correct course. When
shown the survey chart, they indicated with a pencil
which channels led to the sea and which were closed. The
ships set sail, tacking from shore to shore, and the
Indians, unfamiliar with the method of tacking, kept
arching their arms above their heads to indicate that
Frederick Arm was closed and would not take the
schooners to the sea.

The Spaniards proceeded to the Canal de los Nodales,
one of the few names bestowed by them which has been
retained in this area. They anchored there between
Thurlow and Sonora Islands and sent Salamanca with a
party in the longboat to explore the channels to the north,
as Galiano said the information received by the Indians
differed from that provided by the English in Johnstone's
chart.

Salamanca sailed up Loughborough Inlet, which

Galiano named *Brazo de Salamanca*, finding that one river leading off it was blocked by a fish weir. Indians from a village here came out in two canoes, shouting and holding up an otter skin, but the Spaniards were forced to cross to the opposite shore to avoid being hit broadside by the heavy seas. Taking this as an unfriendly gesture, the militant Kwakiutl Indians went ashore, donned cuirasses of hide and pursued the longboat. Salamanca "took suitable measures for using in his defense the slight means at his command", which sent the Indians ashore, but they pursued the boat for some distance on foot, holding up otter skins and arrows to trade.

Upon Salamanca's return, the schooners re-entered Cordero Channel and circled East Thurlow, coming down Mayne Pass (which they named *Canal de Olavide*). Here the force of the eddies and whirlpools threw the *Mexicana* with such violence against the northeast shore that it was feared she would be dashed to pieces. Fortunately, the current carried her off again, and the two schooners continued along to Chancellor Channel (the *Ensenada de Viana*) where they anchored. It had rained steadily all day, and there were squalls in the night, but they were cheered to note that the winds were shifting to southeast which would favour their course.

*   *   *

Commander James C. Prevost anchored his paddle-sloop *Virago* in Knox Bay, West Thurlow Island, in June of 1853 and named the bay after his sub-lieutenant, Henry Needham Knox, R.N., who surveyed the bay during that month. Captain Richards named Chancellor Channel, north of West Thurlow, in 1860. He chose the name because the channel adjoins West Thurlow Island and Loughborough Inlet, both named by Captain Vancouver after Lord Chancellors of England. Richards also named Mayne Passage between the Thurlows after

Lieutenant Richard Charles Mayne, R.N., who served
on the surveying vessels *Plumper* and *Hecate*.

Captain Pender in the *Beaver* named Hemming Bay on
East Thurlow after Pinhorn L. Hemming, a draughts-
man under Captain Richards. A small islet in the bay is
named Pinhorn. Greene Point, on the northeast tip of
West Thurlow, was named by Pender for Lieutenant
Molesworth Greene Jackson, R.N., of H.M.S. *Topaze*.
Off this point are the Greene Point Rapids.

Shoal Bay on the northern shore of West Thurlow was
primarily a logging and mining settlement, and early di-
rectories and Voters Lists show how it was monopolized
by these two industries. One hotel and store were owned
by the mining company and both bore the names "The
Gold Fields of British Columbia." The 1898 Voters List
gives David Cook, postmaster; Charles Collum, hotel-
keeper (Hotel Thurlow); Charles Casswell, hotel cook,
and also names an assayer, a logger and seven miners. No
other occupations are listed. Gerald Rushton quotes a
passenger on the *Comox* in 1897 as meeting at Shoal Bay
"Postmaster Neville Smith, and Archibald and John
Collum, proprietors of the Hotel Thurlow, which is doing
a rushing business."[2]

Gold Fields of British Columbia, with head offices in
London, was a large organization with a registered
capital of $3,000,000. Its impressive Board of Directors
included the Earl of Essex and the Earl of Huntington.
Douglas Pine was in Shoal Bay in 1898, working on the
mountain ridge above the bay, bringing ore by pack mule
down to barges at the water's edge. Iron, copper and gold
were being mined.

The Provincial Minister of Mines paid a visit to Shoal
Bay in 1896 and reported: "Shoal Bay is the only attempt
at a town in the district and consists of a moderate accommodation and a few houses, and is a centre of supply
for the mining and lumber camps for a few miles around.

Superintendent Frederick S. Hussey of
the Provincial Police.
*(Courtesy of the Provincial Archives,
Victoria, BC.)*

Mike Manson. *(Courtesy of the
Ellingsens.)*

Read Island postmaster Lambert and
family, c. 1925. Note method of
rowing. *(Courtesy of Bill Whittington.)*

Logger William Brockman, who was
Read Island postmaster in 1922.
(Courtesy of Bill Whittington.)

Wylie's Hotel, Read Island, c. 1906. Extreme right: the Alfred Joyce family, with Arthur Valdes Joyce, first boy born on Quadra Island. On steps, back; Joe Bigold; Charles Hall. front: Frank Gagne; Joe Silva. *(Courtesy of the Campbell River and District Museum.)*

Barefoot Read Island schoolchildren, Burdwood Bay, 1921.
Back row, l to r: Teacher Ruth Essler; James Lambert; Milton Lilja; Bill Whittington; Forest Lambert; Dwayne Landers.
Front, l to r: Lida Whittington; Bonnie Whittington; Sylvia Landers.
(Courtesy of Bill Whittington.)

Read Island picnic, 1926. Rev. Alan Greene, right front.
(Courtesy of Doris Davies.)

First schoolhouse at Surge Narrows, Read Island, c. 1928. The Tiptons' two bulls hauled the logs and the school was built by the local residents.
(Courtesy of the Vancouver Public Library.)

The steamer *Chelohsin* docking at Surge Narrows, Read Island, c. 1940.
(Courtesy of the Redfords.)

Tipton's store, Surge Narrows, Read Island, c. 1942.
l to r: R.S. Tipton; Clarence Keeling; Nell Tipton.
(Courtesy of the Campbell River and District Museum.)

Deceit Bay (now Redonda Bay) cannery, c. 1915.
(Courtesy of the Provincial Archives, Victoria, BC.)

Lacking gravel, Gerry Olmstead built logging roads of planks.
Redonda Bay, c. 1938.
(Courtesy of the Olmsteads.)

Village of the Friendly Indians, Stuart Island. This is reproduced from
Vancouver's *Voyage of Discovery*, 1798, from a sketch done on the
spot by T. Heddington.

The Yaculta Rapids seen from Big Bay, Stuart Island, c. 1948. Left and
centre are Big and Little Gillard Islands.
(Courtesy of the Brimacombes.)

Maurelle Island school, 1929. *(Courtesy of the Redfords.)*

Picnic on the *Rendezvous*, Okisollo Channel, 1937. Rev Alan Greene
second from left. Maurelle Island in background.
(Courtesy of the Clements.)

Niemi Logging Co. Bunkhouse, Maurelle Island, 1936.
(Courtesy of the Redfords.)

Niemi Logging Co. truck, Maurelle Island, 1936. Note the solid tires.
The boat was used to transport crews to logging locations.
(Courtesy of the Redfords.)

Owen Bay, Sonora Island, 1926. The Logan Schibler house in the background was moved here on a float. The tub was for soaking fishing nets to kill bacteria.
(Courtesy of the Clements.)

Marion and Pearl Schnarr with their tame cougar, Sonora Island, c. 1936. *(Courtesy of the Clements.)*

Steam donkey, Skookum Logging Co., Okisollo Channel, Sonora Island, 1942. Jimmy Van der Est in foreground. Shortly after this, a fire destroyed 3000 logs, including a "cold deck" pile, and the steam and gas donkeys.
(Courtesy of the Redfords.)

Logging at Thurlow Island, c. 1900.
(Courtesy of the Provincial Archives, Victoria, BC.)

Loggers posing on board the old paddlewheel steamer *Beaver* at Thurlow Island, shortly before the *Beaver* was wrecked on Prospect Point at the entrance to Vancouver harbour, 1888.
(Courtesy of the Maritime Museum, Vancouver.)

The historic tug *Active* leaving Forward Bay, Cracroft Island, with A & M Logging Co. boom, 1947. The *Active* was built in 1889; in 1952 she was rebuilt into a diesel towboat. Photo by O.F. Landauer.

Loading truck at Forward Bay, Cracroft Island, 1947. Note solid tires and plank road. Photo by O.F. Landauer.

Tenaktak House, Harbledown Island, scene of a Kwakiutl legend. Photo by E.S. Curtis, 1914.

The Reverend Alfred J. Hall, pastor and teacher at Alert Bay, Cormorant Island, and Mrs. Hall, c. 1876.
(Courtesy of the Provincial Archives, Victoria, BC.)

The old Indian residential school at Alert Bay, established in 1894.
(Courtesy of the Provincial Archives, Victoria, BC.)

Potlatch at Alert Bay, c. 1910. Photo by William Halliday.
(Courtesy of the Vancouver Public Library.)

The boardwalk in front of the Indian community houses, Alert Bay, c.
1917. *(Courtesy of the Provincial Archives, Victoria, BC.)*

Alert Bay School, 1911–12.
Back row, l to r: Bobby Hunt; George Robinson; Bob Spouse; Ethel Halliday; Bertha DeRooter; Milton Adams; Herbert Davis.
Second row, l to r;: Diana DeRooter; Fred Wastell; Miss Tarbuck, teacher; Lenora Adams; Irene Fleming.
Front row, l to r: Leonard Hunt; Wilhemina Davis; Rosie Fleming; Ollie Huson; Muriel Fleming;Robert Davis.
(Picture and names courtesy of R.W. Davis, Sr.)

B.C. Packers' cannery, Alert Bay, c. 1930.
(Courtesy of the Provincial Archives, Victoria, BC.)

Finnish settlers at Sointula, Malcolm Island, 1909. Photo by Andrew Anderson. *(Courtesy of the Provincial Archives, Victoria, BC.)*

The Finnish settlement at Sointula in earlier days. The large white meeting house, far left, still stands today.
(Courtesy of the Provincial Archives, Victoria, BC.)

The hotels were full to overflowing at the time of my visit, and I would have been at a loss to find accommodation, but for the kindness of Mr. E. Pooke, agent of the Gold Fields of B.C., who kindly placed at my disposal an unoccupied house belonging to his company."³

About a mile west of Shoal Bay is Bickley Bay. Sven Hans Hansen ran a store there in the early days, along with Nels Hjorth. The two Norwegians came to Canada as crew on a merchant ship, found conditions aboard were not to their liking and jumped ship near White Rock. Hansen became well known along the coast as Hans the Boatman. He lost his hand in an accident and had it replaced by a hook that was shaped to hold an oar. He ran his small boat skilfully, using a sail and one oar. In 1891 he moved to Port Neville and in 1895 became the first postmaster there.

Logging camps opened up on the Thurlows in the 1880's. It was hand logging at first, using oxen teams, and the remains of some of the old skid roads can still be traced at Shoal Bay. The old paddle-steamer *Beaver* was sold by the Hudson's Bay Company in 1874 to Stafford, Saunders, Morton and Company of Victoria, who altered her into a general freight and tow boat to supply the logging camps before the Union Steamships started their service. She was setting out on one of these trips when she was carried by the tidal current on to the rocks at the foot of Prospect Point bluff in Vancouver Harbour and wrecked. It was a dark night, on July 26, 1888. The assistant engineer, W.H. Evans, gave this account to Major Matthews, former Vancouver archivist: "She had been running north, and this time, the night she was wrecked, it was dark, about one a.m. in the morning, we were going to Nanaimo for bunker coal before going north to some island, Harwood or Thurlow."⁴

When the Union Steamship *Comox* took over, she used to stop in midstream along Cordero Channel and hand loggers rowed over from sites around Blind Channel and Greene Point on the Thurlows to collect letters and supplies. Shoal Bay became the regular stop, and settlers would row over from the surrounding islands to get their mail.

It has been said that the first steam locomotive used for logging in British Columbia was on Thurlow Island. Sol Reamy was hauling logs with the famous little locomotive "Curly" for the Hastings Sawmill Company on Thurlow and Quadra around 1901. John Antle reports the Hastings Sawmill Company in operation again on Thurlow in 1906, and a new mill opened by Thurlow Island Lumber Company at Blind Channel in 1907, with Bert Campbell as superintendent. The mill employed a large number of men for several years sawing cedar for a Vancouver factory. At Greene Point, a logging company was operated by A.P. Allison.

A one-room school opened at Shoal Bay in 1913, with teacher Miss Lilian Hood and trustees A. Prichard, secretary P. McDonald and J. McKelvie. The government built a trail from Bickley Bay to Shoal Bay so pupils could walk from Bickley Bay to the school.

A government wharf was also built at Shoal Bay at this time. Before World War One a telephone line was laid, connecting Shoal Bay with the Forestry station at Thurston Bay on Sonora and with Rock Bay on Vancouver Island. A government surveyor came up on the *Cheakamus* in 1915 and laid out lots on the north side of Hemming Lake on East Thurlow. He noted the land had been logged over some 22 years previously and felt it would make good farmland. Despite its predominantly rugged terrain, there has been some farming on West Thurlow at scattered points, with vegetables the chief product.

The *British Columbia Directory* of 1918 gives the population of Thurlow as 50, and the industries as lumbering, mining and fishing. Although according to the 1919 Mining Report there was no activity on Thurlow or in the neighbourhood from 1902 to 1919, there were still several mining companies listed in the 1918 directory for Thurlow: Monte Cristo Mining, Alexandra Mining and Douglas Pine Mining. Lumbering companies listed were McKee and Milens Lumbering, O'Connor Logging, Hastings Sawmills and A.P. Albron Lumbering. Seven lumbering companies are listed for 1921. Peter McDonald had the Shoal Bay store and post office in 1918 and into the 1920's.

For some years there was a cannery on the east shore of West Thurlow, at Blind Channel. W. Anderson and Co., fish canners, sometimes referred to as the Quathiaski Canning Company, is listed in the 1918 directory and continuously through 1936. Frank Allen was manager in 1923 and Boecher Hall in 1928; both also served as postmasters at Blind Channel. The cannery building was torn down in 1976. Mr. and Mrs. Richter run the store, post office and liquor outlet today and provide showers and a laundromat for tourists.

Thurlow Island's logger story of the 1920's concerns the burial of an old logger named Brown, who was drowned in Phillips Arm and washed ashore much later in Bickley Bay. Police came from Campbell River for the inquest at Shoal Bay, first inspecting the body and then requesting Brown's logger friends to build a coffin and bury him. They agreed to do this, but went first to Shoal Bay to fortify themselves for the ordeal at the hotel bar. Logger Ed Dolby finally remembered his old friend lying on the shore at Bickley Bay and rowed down in the dark. He constructed a rough coffin, got Brown's remains into it and nailed down the lid. It had been a tiring and unpleasant task, and by the time the other loggers missed Dolby and rowed down after him, they found him lying asleep in the newly-dug grave, clasping a bottle to his chest. Abandoning the first impulse for a practical joke which the situation suggested, they woke him to help them resolve a problem that was puzzling them.

They all firmly believed that to ensure repose for the dead, the body must be laid to rest with feet pointing towards the dawn. Opening the coffin appealed to none of them. Dolby solved the dilemma. "Tilt the coffin," he said. "He's wearing his corked boots; whichever end makes the loudest thump, that's his feet." Solemnly, by the light of an old Cold Blast lantern, they tilted the coffin first one way, then another, listening intently. Brown's feet were finally located, and he was placed in his grave, feet towards the dawn; he was assured now, his friends believed, of restful slumber throughout eternity.

The original hotel and stores are no longer at Shoal Bay, and except for a family sawmill just south of the bay there is little industrial activity. There is a long government wharf to serve boats that anchor in the bay, but no settlement remains. The streets that were laid out and named, ready for the expected boom, now contain only a few, deserted, ramshackle houses, though Hank and Camille Dirksen run a roomy lodge that extends over several of the original lots. It has a large living room with an attractive stone fireplace and offers year-round service to the many fishboats and pleasure craft that wait in Shoal Bay before entering Greene Point or Yaculta Rapids.

Not long ago, Phillips Arm, across from Shoal Bay, attracted numerous tourists with its big salmon runs up the Phillips River to beautiful Phillips Lake, but floods altered the river course and comparatively few salmon are caught there today. However, during the summer season, an influx of tourists monopolizes the government wharf at Shoal Bay and commercial fishermen are forced to anchor out in the bay. With the arrival of fall the tourists depart and the commercial fishermen take over the wharf once again.

CHAPTER 9

Hardwicke Island

This triangular island of 27 square miles, with its base lying along Johnstone Strait, was named by Captain Vancouver after Philip Yorke, third Earl of Hardwicke, grandson of the first earl who had been Lord Chancellor in 1737. Philip Yorke was Lord Lieutenant of the county of Cambridge and, as he was a patron of Spelman Swaine, R.N., of the *Discovery*, who came from that county, Vancouver wrote in his journal that the naming of the island was in compliment to Swaine. Yorke Island, off the western tip of Hardwicke, was named after the Hardwicke family by Captain Richards when he was surveying in the *Hecate* in 1862. Mount Royston, 2625 feet high, on Hardwicke takes its name from Royston in the county of Cambridge, where Philip Yorke was a member of Parliament, from his graduation from Queen's College, Cambridge, until his accession to the peerage.

Wellbore and Sunderland Channels, faced by the other two sides of the Hardwicke triangle, were named by Cap-

tain Pender in the *Beaver* in 1863, both after the
Honourable John Wellbore Sunderland Spencer, R.N.,
captain of the H.M.S. *Topaze*. Topaze Harbour, above
Sunderland Channel, was named by Pender after the
51-gun *Topaze*. It was from Jackson Bay in Topaze Harbour that the Lekwiltok branch of the Kwakiutl
descended upon Quadra Island, driving out the Salish.
James Johnstone, R.N., master of the *Chatham*, after
whom Vancouver named the strait, passed by Hardwicke
early in July 1792, seeking a passage to the Pacific. When
he returned to report the opening, Vancouver and his
party sailed up Discovery Passage, past the Thurlows and
Hardwicke to the open sea, and south to Nootka.

On July 27, 1792, at seven A.M. the Spanish ships *Sutil*
and *Mexicana* sailed to the south of Wellbore Channel,
which Galiano named *Canal de Nuevos Remolinos*. The
longboat was manned and sent ahead to guide them
through the channel. They reached an anchorage which
they named *Novales*, on the northwest side of Hardwicke
Island, and remained there while Valdes went exploring
in the longboat from July 31 to August 7.

*   *   *

Two logging camps, Anderson's and Paterson's, were
active on Hardwicke in 1916, according to John Antle,
who said Paterson's camp number 3 was planning to take
out seven or eight million feet. Anderson's camp had built
a number of houses so married loggers could live with
their families, and dances were organized to keep them
content on the isolated island. Loggers Wertenan and
Jetter were closing their camp at this time, having taken
out all their logs from the island.

A man named Cook is said to have logged at several locations on Hardwicke with a team of oxen in the early
1900's, leaving behind him a few deserted buildings near
Earl's Ledge, a reef that juts out into Johnstone Strait on

the south shore of Hardwicke. Here, summer westerlies as well as southeasters are winds to fear when they are combined with the swift tides of Current Passage that swirl and boil over Earl's Ledge. The section of Johnstone Strait that runs past Hardwicke can be one of the roughest passages along the inside route at such times. Then again, on a calm morning, the sea lies quiet and serene. The area is ideal for logging and fishing, and inspired the first settler, William Kelsey, to pre-empt 114 acres in 1911. Kelsey, a logger, came with his wife and children from Topaze Harbour and after the move he continued to hand log there and on Hardwicke and Vancouver Islands.

Around 1911, Japanese loggers came to the island with horses and, later, with steam donkeys. Settlers remember that the loggers treated their huge horses with great solicitude. During the New Year celebrations of the Japanese, a good-luck food stack of rice cakes and oranges, topped with a lighted candle, was placed before the stall of each of the horses. If the candle burned down, good luck followed, but if it went out, ill luck during the coming year was predicted for the unfortunate animal.

In 1916 the Minister of Lands reported pre-emptors living in cabins on Hardwicke, with some clearing done. Early settlers and loggers included such names as Edwards, Thompson, Erickson, Forberg and Hand.

In 1918 a Norwegian, H.A. Bendickson, moved to Hardwicke Island with his family. Bendickson had logged since 1904, first hand logging, then using a steam donkey, and finally founding a large logging and towing firm. He logged on various islands, including Quadra and Read, and was logging in Jervis Inlet when he decided to locate his home on Hardwicke.

The Bendickson tug was away, hauling a log boom, so a small, coal-burning, steam tug, the *Rex*, was engaged to tow the large boom float on which the Bendickson house, camp building and logging equipment were assembled. It would be an eight-day trip through a succession of rapids to reach Hardwicke. The tug captain looked at the boom float with its widely-spaced, slippery logs, and then at the seven lively young Bendickson children ranging in age from two to 12, and refused to take on the job. Finally he was persuaded when he was told the teacher would be travelling with the family and the children would be kept corralled in the house, studying their lessons.

The weather was fine and the sea calm as they sailed down Jervis Inlet, and the older boys managed to elude their teacher and slip out unobserved to catch a few salmon off the rear of the boom. When the Yacultas were sighted the captain took no chances. He crowded the Bendickson family, the teacher and the logging crew into his tugboat, while the mate was stationed by the tow bit with an axe poised to cut the tow line if the boom was smashed against the shore. All went well until they reached Shoal Bay on East Thurlow Island. Here the tug ran out of coal and there was an enforced halt as the crew went ashore to saw logs on the beach and get the steam up again with firewood.

A schoolhouse was promptly built by Bendickson and Kelsey. It also served as a honeymoon cottage for a member of the logging crew. During the move to Hardwicke, one man had asked to be put ashore at Powell River to answer a call of nature. There was no sign of him when the tug was ready to proceed, and it set off without him. The man, it seems, had answered a different call of nature. He had gone down to Victoria, got married, and eventually turned up at the Hardwicke camp accompanied by his bride. The Bendicksons, respecting a need for privacy, let the couple sleep in the schoolhouse at night, and teacher and pupils took over during the day.

Kelsey moved across the strait to Kelsey Bay in 1922

and his property was acquired by Bendickson. For many years the Bendicksons logged under a Tree Farm licence, reseeding or planting after cutting the trees. Tree Farm licences were first issued in 1951 for Crown Granted land under a special tax arrangement to encourage the dedication of suitable private land to permanent forestry, the owner to grow and sell timber continuously on the land. In 1971, the Bendicksons sold their licence to Crown Zellerbach, but Bendickson sons and a nephew bought the logging equipment and still log on Hardwicke, on land designated by the big company or by the Forestry Department. They are the only loggers working on Hardwicke at present.

Alan Greene and his mission boat were always welcome in the early days when they called at isolated islands and Port Neville. Edith Bendickson, daughter of Sven Hans Hansen, pioneer postmaster of Port Neville, has written news articles on those days and remembers her excitement as a child in the 1920's when she saw her first Charlie Chaplin movie, run off by Greene on a little, hand-cranked machine.

Greene was acting minister in 1937 at the wedding of Lily Bendickson to Edith's brother, Olaf Hansen. Over 100 guests came to the island wedding, defying a westerly that whipped the waters of Johnstone Strait. Many of the guests to this Norwegian festivity were ferried across from Kelsey Bay by Bendickson, who bucked tides and currents in his launch. There was dancing to an accordion in the little schoolhouse on the hill, and when the floor-thumping polkas ended at midnight, a lavish Scandinavian supper was served in the house. After that, Bendickson once again ferried boat-load after boat-load of guests at two in the morning through the rough waters back to Kelsey Bay.

It is difficult to dissociate the Bendicksons and Hardwicke Island. Bendicksons have been a part of the island for 60 years. Four of the six sons and two married grand-sons have their own large homes and orchards on Hard-wicke, and it appears that Bendicksons will be on the island for many years to come. They have set up a tele-phone system between their homes and, through their camp office, can contact Kelsey Bay by radio phone.

Commercial and sports fishing figure largely today, and the annual Sayward-Kelsey Bay Fishing Derby is a popular event. Only a few people, mostly Bendicksons, live the year round on the island, but those with vacation cottages increase the population during the summer months. Logging and timber-thinning crews also come and go. For a time there were 16 to 18 families, with 29 children in the school, but the little building is no longer in use. The nearest school now is at Sayward on Van-couver Island, and only one small Bendickson child travels from Hardwicke to attend, taken there by his mother, first by boat and then by car, unless storms whip the seas and keep them both at home.

CHAPTER 10

Cracroft, Minstrel and Harbledown Islands

Beyond Hardwicke, past Port Neville, is larger, sparsely-settled Cracroft Island. Its deep bay, Port Harvey, was called *Insulto* by the Spanish explorers after crew members gathering wood on its shores were attacked by Kwakiutl Indians demanding Spanish muskets. The crew belaboured the Indians with their swords and forced them to retreat to their village. To reassure the Spanish viceroy that his orders to treat the natives "with compassion " were obeyed, Galiano explained that the men kept their swords in their scabbards while using them to beat off their attackers.

*   *   *

Cracroft was named by Captain Richards in 1861 after Sophia Cracroft, niece of the English explorer, Sir John Franklin, who was lost in the Arctic in 1845 while searching for a northwest passage. Burial Cove on the east coast of Cracroft was so called because Indian burial boxes were placed in the trees that lined the bay.

In 1906 the Merrifields ran the store at Forward Bay, where there was also a logging camp. That same year, the Columbia Coast Mission ship *Columbia* reported rescuing part of the camp's log boom that had broken up in a southeaster, and towing it to sheltered Port Harvey. There were three Cracroft logging camps listed as regular stops for the *Columbia* in 1907. Cracroft also had a cannery running in the 1930's, in Bones Bay. Charlie Watson, who was in charge of maintenance and security for the cannery, owned a steam-driven launch, the *Flora Belle*, the last on the coast. His wife wrote a poem for the *Log*, giving her version of the origin of Bones Bay's name. A part of it reads:

> "But time, and rains, did not respect the secret of the past—
> And grinning skulls, and scattered bones, are all revealed at last!"[1]

In the 1940's there was a demand for lumber when building was reactivated after the war. Logging camps sprang up on many of the islands. Forward Bay on Cracroft was the scene of a large logging operation, the A & M, owned by a Texan named Airsley and managed by a Mr. Anderson. The loggers constructed a long, wooden-plank road running around one side of the bay and partway up the mountain. There were more than a dozen buildings, a well-equipped cookhouse and dining room and a sturdy company wharf where the *Cardena* called weekly. In the late 1940's, Canadian Forest Products took over the company and a few years later a disastrous fire burned out most of Cracroft Island, including the camp and all its equipment. Loggers fled down the mountain-side to the sea to escape the swift advance of the flames, and were rescued by a tug whose men had seen the blaze from Englewood in Beaver Cove on Vancouver Island and sped to their aid.

Port Harvey has a look of desertion today, even on the

Indian reserve. In the past, abundant wild berries and seafood satisfied the needs of the natives, but now both Indians and whites have moved to centres offering schools and higher-paying work. The trend of the islands towards the tourist trade is marked here by a marina on the east coast, with seaplane service available.

On little Minstrel Island, off the northeast coast of Cracroft, Oscar Soderman hand logged and built himself a shack around 1905. The hotel and store there were run by Armstrong and Bennet. Tom Bennet, a genial logger and jack-of-all-trades, later bought the store in Port Harvey on Cracroft, where he also served islanders as amateur doctor, lawyer, mechanic, and letter writer for illiterates. Minstrel Island received its mail and that of the surrounding islands when the *Cassiar* called at its post office every two weeks. A minstrel ship from the United States is said to have called there once, and a government survey crew gave Minstrel Island its name as a result. Similarly, Sambo Point on Cracroft and nearby Bones Island were named by the crew after well-known minstrel characters.

The beer parlour of the Minstrel Hotel was the popular gathering place for Cracroft loggers to unwind after a hard week of work, and there are many stories of the high jinks that went on there. In 1907 it was reported in the Vancouver *World* that Port Harvey loggers who celebrated Christmas Eve in the Minstrel Hotel emerged to find that their tug *Leonora* had vanished and was presumed to have foundered and sunk while at anchor in heavy seas. She was seen a few days later at work as usual and no more was said about her mysterious "disappearance." Minstrel Island had a school in 1944, with Mr. Herbeston its first teacher. There is a small schoolhouse there today. Radio telephone service came in 1950. The island still has its post office, radio phone service in the hotel, a liquor store and two resorts offering year-round service to sports fishermen, with flights available to Campbell River.

The next island along Johnstone Strait is heavily-forested Harbledown Island. Like Cracroft, it is largely timber-leased land, its few pre-empted lots clustered along the centre of the northern coast and small Indian reserves located at the northern tip and on the southwest peninsula south of Parson Bay. E.S. Curtis has written of the Kwakiutl tribe, Awaitlala, in the northern winter village, whose name means "eelgrass on point." He photographed a house in the village, Tenaktak House, the home of a young Indian who, according to legend, slew a mythical monster that stole salmon from the Awaitlala.[2]

An Hawaiian, Kamano, married to an Indian woman, was the first settler on Harbledown, and tiny Kamano Island off the east coast bears his name. Kamano's two daughters married early loggers: Olney, who logged with the first oxen team on the island, and Jolliffe, who owned a steam tug to tow the logs. William Galley pre-empted land after this and served as storekeeper and postmaster. He died in St. George's Hospital in Alert Bay in 1926. There was no float when Galley was postmaster; mail was transferred in midstream from steamer to rowboat or left at Minstrel Island. Despite its sparse settlement, Harbledown had a school in 1910, the teacher being Miss Monroe. The island is still heavily timbered and logging is going on. Beware Passage, Care Island and Caution Rock along the northern coast are labels to warn sailors of dangerous rocks and currents around this green, forested island.

CHAPTER 11

Cormorant Island

"It is a beautiful sail from Alert Bay to Kingcome
Inlet—a perfect archipelago, reminding one of the
Thousand Islands in the St. Lawrence."
                    -A.F. Cotton, 1895[1]

Past the archipelago of islands that fringes the eastern
waters of Broughton and Queen Charlotte Straits lies
little, crescent-shaped Cormorant Island, site of the
village of Alert Bay. Home of one of the most powerful of
the Kwakiutl tribes, and lying directly opposite the
Nimpkish River that teems with sockeye salmon during
the fish runs, Alert Bay is an important fishing area with
a stormy history involving whites and Indians.

Both Cormorant Island and Alert Bay were named
after armed vessels; Cormorant Island was named by
Commander George T. Gordon in 1846 after his ship,
H.M. paddle-sloop *Cormorant*, six guns, Alert Bay by
Captain Richards in 1860 after H.M. screw corvette
Alert, 17 guns.

The huge sockeye runs were responsible for the first

settlement on the island. S.A. Spencer, originally a Victoria photographer on Fort Street, with W. Huson, who had managed a coal mine at Fort Rupert, decided to establish a cannery in this productive area. In 1865 they came up by canoe; they went first to the Nimpkish Indian village, but found the water too shallow in the river mouth. They leased Cormorant Island from the government instead, and in August of 1870 brought up lumber and supplies on the *Emma* to build a wharf, store and small saltery for curing salmon.

Nimpkish Indians were employed in the saltery, but showed little interest in a steady work week. Most of their needs were supplied by the abundance of fish in river and sea, and they took off for their homes by the Nimpkish River whenever the mood struck them. Alert Bay, which they called *I-lis*, meaning "Spread Legs Beach", had been an area where various Kwakiutl tribes congregated for winter potlatches, returning afterwards to their villages. The saltery owners decided the only solution was to locate Indian workers permanently in Alert Bay.

Alfred James Hall of the Church Missionary Society was running a mission in Fort Rupert, near present-day Port Hardy on Vancouver Island. Spencer and Huson persuaded him that a more convenient site would be Alert Bay, on the direct route for passing ships. As a further encouragement, they built him a large house. In 1878 Hall moved across the water to Alert Bay, bringing his pupils with him and followed by many of their families. There were 16 houses in the Indian village in Hall's day, containing 83 men and 79 women. Hall altered his house to make it suitable for a small residential school, which opened with six pupils. Mrs. Hall taught the mission girls housekeeping, music and English, and six years after the move, Mr. Hall built a sawmill with funds from the missionary society so the boys could be taught a trade. The Halls lived there for 32 years, during which time Hall translated the Bible and various hymns into the Kwakiutl language. Mrs. Hall, homesick for England in this alien atmosphere, planted foxgloves on the hills and in the clearings to remind her of her native land.

Indian Agent Pidcock had his headquarters originally in Fort Rupert but when fire destroyed the fort, it was decided to move the agency to Alert Bay, 20 miles away by water. Mrs. Pidcock and four of her children stayed on in Fort Rupert until Pidcock could build a house large enough for their big family: Pidcock and three of his boys camped in a tent at Alert Bay. "Wet weather," he wrote in his diary on August 25, 1888, "very unpleasant especially with three boys in a tent." A delegation of Indians greeted him by announcing they resented his housebuilding and would leave the island if he went on with it. "I merely said that I was ordered to come there and I obeyed my orders, and I hoped they would say nothing more as it was foolish."[2]

The Kwakiutl had stated their position clearly in the 1880's when anthropologist Franz Boas visited Alert Bay. "It is a strict law that bids us dance," they told him. "It is a strict law that bids us distribute our property among our friends and neighbours. It is a good law. Let the white man observe his law, we shall observe ours. And now if you are come to forbid us to dance, begone, if not you will be welcome to us."[3]

There was no love lost between Indians and the whites who sought to enforce their own laws. Around this time, there was a report of Indians drinking on the reserve. Constable Woollacott and an assistant entered one of the houses and handcuffed the offenders, whereupon their friends stunned the assistant with a loose doorknob and threw Woollacott bodily through a closed window, his face and hands suffering cuts from the broken glass. Special constables were sworn in, and about ten Indians were arrested.

There were sympathetic white residents who encouraged the Indians and aided them in preparing appeals against potlatch convictions. "The more I see of this place the less I like it, I mean some of the People in it," said Pidcock, lonely for his family, working every day at his housebuilding and spending uncomfortable nights in his wet tent. But he found the Halls "very kind." Mr. Hall lent him a small canoe in which he set forth, alone, up the stormy coast to visit his wife, Alice, in Fort Rupert, arriving at six P.M. in the dark, just ahead of a southeaster. When it had subsided somewhat, he started the 20-mile return trip, paddling hard all the way, but making two overnight stops, one at a small Indian house, "horribly smoky."[4]

Moving day came at last on September 15, and the family left Fort Rupert in three loaded canoes, accompanied by four Indians. When they were halfway down the coast, a gale blew up and they were forced ashore at False Head, taking shelter in two tents which they had with them. The storm continued unabated, and they despatched one of the Indians the next day to find a cabin and bring back food. He returned with a few potatoes and one loaf of bread, soon demolished by their large party. There was no let-up in the storm, and the Indian was sent off once more. He returned on September 21, this time with enough food for one good dinner for all, which they devoured ravenously. There was no need for rationing, since the worst of the storm was over. They left False Head the next morning, after seven unpleasant days of cold, damp and hunger.

During his stint at Alert Bay, Pidcock tried to obey the difficult directions of the Indian Affairs Department to civilize the Indians, abolish the potlatch and refrain from antagonizing the natives. According to William Halliday (one of his successors as agent), Pidcock felt strongly that the residential school was the best means of training young Indians, the only way to circumvent the antagonism of their parents towards the white culture. In Pidcock's 1888 report to the Department of Indian Affairs he says: ". . . the children are not averse to learning but their parents see in education the downfall of all their most cherished customs."[5] After a lengthy correspondence, he succeeded in having about 450 acres of land on Cormorant Island set aside for school purposes, and here a residential school for Indian boys was built in 1894. Mr. A.W. Corker was the first principal. William Halliday came in 1897 to assist Corker for a few years before becoming Indian Agent. He was agent for 38 years, describing many of his experiences in his book *Potlatch and Totem.*

Indian parents during the first years of the school were opposed to the isolation that weaned their children from Indian ways, and the boys themselves were rebellious. The 30 boys, natural-born fishermen, were taught carpentry, furniture making and boatbuilding, as well as cattle and horse raising, none of which appealed to them. Many grew homesick. One boy slashed himself with a knife and said he had been attacked in the woods, hoping this would allow him to be sent home, but Corker bandaged the cuts and the boy was kept in school. Fires were also started in the school building by several of the homesick children.

Early pioneers were Stephen Cook and his wife. Cook was born in Victoria of a white father and an Indian mother, and at age 11 was sent to the Metlakatla school, then to the Fort Rupert Mission, and finally to Alert Bay, where he became book-keeper and secretary to Mr. Hall. A young Indian girl had been left in the care of Mrs. Hall as a child by her father who was travelling to Alaska in a small boat and never returned. In 1891 she married Stephen Cook who by then had been put in charge of the mission sawmill and store. When the store was sold to British Columbia Packing and Fishing Company, Cook built and ran a store of his own. Sons of the Cooks became fishermen, one daughter was a nurse and another became principal of a public school.

Around 1881, S.A. Spencer took Thomas Earle as a partner and the two built a cannery on the wharf and ran the post office from their store. H.J. Young was the cannery superintendent. B.C. Packers bought them out in 1902, sold to B.C. Packing and Fishing Company in 1909, and bought the cannery back again in 1928. After the Canadian Pacific Railway was built in 1886, Alert Bay was kept busy supplying the rapidly-growing populations of Victoria and Vancouver with fish and lumber.

There were very few totem poles in the Indian village before 1880; it was about 1890 that the elaborate poles began to appear. Around this time Philip Woollacott was constable for Alert Bay, with several Indians sworn in as assistant constables. An English settler, Granville Lansdowne, shot three Indian dogs that he said had been worrying the cattle on his Knight Inlet farm, and the Indians were enraged, as these were valuable bear dogs. After a fracas between settlers and Indians, two Indian constables arrested Lansdowne and brought him to Woollacott in Alert Bay. To their amazement and chagrin, Woollacott told them they, and not Lansdowne, were at fault, since they had made an arrest without a warrant, and in any case had no authority off the reserve. The Indian constables were put under arrest until Agent Pidcock returned to the bay. He dismissed them from the force and told them to convey Lansdowne back to his home. The Indians loaned the settler a canoe but refused to accompany him after a judgment which to them was unfair and incomprehensible.

Property had been set aside in 1899 for a school for white children on land along the waterfront. The Public Schools Report shows a school in 1900, with P. Woollacott as teacher with a salary of $50 a month. Woollacott children supplied five of the 15 pupils. Six of the remainder were children of Silas Olney. During the first year of the school, Spencer estimated the population of the island at about 60, mostly Indian. There were no roads, and travel was by water. Wages at that time were $2.50 a day for whites, $1.50 for Indians.

The Columbia Coast Mission had been serving 84 logging camps with its hospital ship *Columbia* since 1905. When calling at settlements, the mission boat often gathered residents together and carried them to picnics or Christmas parties, with a brief sermon before the entertainment. In 1908 she carried the Alert Bay Glee Club to Scott Cove on Gilford Island, 25 miles away. Arriving at night, the crew found boomsticks stretched across the entrance to the cove, but the enterprising minister sent the ship hopping up and over them, both coming and going. In 1909 the mission opened St. George's Hospital in Alert Bay on a one-acre site donated by B.C. Packers. The following year, a second *Columbia* replaced the first ship; *Columbia II* was later sold to a new owner who christened her *Wayward Lady*.

In 1909, most of the trees had been removed from the island by loggers, and ugly stumps were eyesores on large sections of the island. The settlement consisted of a few buildings in the bay, strung along the waterfront. At night, residents carried lanterns to light their way along the dark trail. The Indian reserve and Indian Agent's house were at the west end; then came the Indian school, the homes of Chambers, the cannery manager and Woollacott, the constable, the public school, the hospital, and finally the homes of Dave Huson, Spence Huson and Silas Olney. Jack Robilliard, a fisherman, lived in a log cabin in the woods, as did George Hawkins, a coal burner, who supplied charcoal to the Chinese in the canneries to use in soldering their cans. These were handmade in the days before the invention of the crimping machine eliminated the need to solder.

George Hawkins was a well-known, long-time resident on the island. Listed as a coal burner there in 1882, he later ran a fruit farm and at one time was a member of the school board. As he aged, his white curls and beard gave him the appearance of a rather rakish Santa Claus. He used to capitalize on this during the Christmas season, calling at many of the homes to tender greetings from Saint Nicholas, and usually receiving an invigorating drink at each one, so that by the end of his tour he resembled Clement Moore's saint even more closely, with "cheeks like roses and nose like a cherry."

A.W. Wastell arrived in Alert Bay in 1909 to take charge of B.C. Packers' sawmill and box factory. His young son Fred, returning home from university in 1919, brought his Chevrolet car with him. The only place he could drive it was on a short stretch of rough road along the waterfront from the Indian Agent's house to the residential school, but he had the distinction of owning the first car on the island.

William Halliday was instrumental in helping a young Chinese family get established in Alert Bay. Jim King arrived in 1910 to work in the sawmill, and when his wife and children came from China to join him he started a small cigar and tobacco shop. After Halliday helped him obtain foreshore property he was able to open up a successful grocery business which he managed until the 1930's, selling out then to Dong Chong.

In 1911-12, a telephone-telegraph service was established between Campbell River and Alert Bay, and in 1919 the North-West Telephone Company's station was moved from Beaver Cove to Alert Bay. The cannery built a big wharf in 1916, close to the Indian village. The sawmill was working steadily and prosperity seemed assured for Alert Bay. For the Indians, things ran less smoothly.

Indian Agent Halliday had been speaking out strongly against the potlatch, saying it made the Indians apathetic, that they were wrapped up in their old customs and lived only for the potlatches and dances for which their big buildings were designed. He brought charges against two Indians of Alert Bay in 1914 and continued his complaints and charges throughout the period of 1914 to 1920, declaring the potlatching increased tremendously during that period. Letters from Indians and their supporters swamped the newspapers and the government; missionaries and teachers took the opposite stand with equal vehemence.

Then came Dan Cranmer's enormous potlatch, the Christmas Tree Potlatch, the biggest ever, given defiantly by the Alert Bay chief in December of 1921 and held on Village Island to escape the watchful eye of the Indian Agent. Betrayed by one of their own race, an Indian constable who helped the dancers with their costuming, the hundreds of Kwakiutl guests from all parts of the Kwawkewlth area were reported to the agent and a great many faced court charges. The schoolhouse in Alert Bay was used as a courthouse, and during the trial period the Indians slept on the school floor at night, guarded by the Royal Canadian Mounted Police.

E.K. De Beck, son of the Indian Agent who preceded Halliday, was one of the defense lawyers. He won an option for the Kwakiutl: jail sentences could be avoided if the accused would agree to give up potlatching and dancing and turn over to the court all their paraphernalia, which would be sent to Canadian museums. Although Alert Bay joined Cape Mudge and Village Island in agreeing to the proposition, potlatching continued there in secret. The curtailed ceremonies lost their original significance, however, and many of the songs and dances were gradually forgotten. Today, with the Potlatch Law erased from the statutes, Alert Bay is attempting to revive some of the dances as a cultural heritage for future generations. Ironically, the watered-down Indian dances and "potlatch" have become regularly-scheduled tourist attractions, advertised by the provincial government.

When the heirlooms of the various families in the three

areas were surrendered, they were sent to Halliday in Alert Bay. He wrote to Ottawa: "I have piled in my woodshed at least 300 cubic feet of potlatch gear.... It will be a very valuable and very rare collection and should command good prices for museum purposes.... Many of the American Museums would simply jump at the chance of obtaining it."[6]

Halliday did sell some to G.W. Heye of a New York museum, in an effort to get as good a price as possible for the Indians. Seventeen remaining cases were shipped to Ottawa, with a list of the articles each family had relinquished, but 44 pieces were found to be missing, stolen *en route*, when the cases arrived in Ottawa.

Now, 50 years later, Ottawa has returned the artifacts and has authorized a large complex at Alert Bay for the Nimpkish paraphernalia. Artifacts of the Village Island families will be displayed at Alert Bay or Cape Mudge, or possibly will rotate between the two museums. (See also the accounts on pages 59–61.)

The first hotel on the island was the Nimpkish, built in 1920 on the foreshore of the Indian reserve and originally owned by the Beswicks. The hotel owners wanted to include a beer parlour on the premises and as Indians were still forbidden to buy liquor, it was decided to move the hotel off the reserve. In 1925 the large building was loaded on a scow, towed to the eastern end of the bay and set up there on the waterfront.

St. Michael's Residential School, with room for 200 Indian children, opened in 1929, replacing the original school started by Pidcock. A huge edifice, it loomed up on the waterfront with its own farmland lying behind it. G.H. Skinner started a dairy farm the year the school opened and A.D. Smith ran a chicken farm; both supplied St. Michael's as well as the logging camps. An effort was also made by the school staff to serve whenever possible a selection of native foods such as seaweed, oolachan grease and the great favourite, soopolallie berries whipped to a froth. A superior school opened in 1939, to which Indians in the residential school could transfer after reaching grade nine. In 1955 a modern school replaced it, integrated at first from grade seven.

Alert Bay Indians have been active in organizations formed to improve working conditions and maintain rights. Chief William Scow was one of the Christmas Tree potlatchers convicted in the 1922 trials. "When you took the potlatch away from us, you gave us nothing to take its place," he told the Indian Affairs Department. He became president of the Native Brotherhood of British Columbia which published the newspaper *The Native Voice*, its object to fight for equal rights for Indians, and he made numerous representations to federal and provincial governments demanding the franchise. When this was granted in British Columbia in 1949, Scow and Frank Assu travelled to Victoria to thank the legislature. Chief Scow was invited to Queen Elizabeth's coronation in Westminster Abbey in 1953 and was received at the palace in full Indian chief regalia, along with Chief Billy Assu of Cape Mudge.

With fishing a major industry of Alert Bay, both its Indian and white fishermen have played a large part in fishing unions and protests. Alert Bay was the scene of organizational meetings bringing the Kwakiutl Indians into the Pacific Coast Native Fishermen's Organization, which later amalgamated with the Native Brotherhood. It was in Alert Bay that the famous strike of 1938, with the flotilla descent from Quadra Island to Vancouver, was organized by the white fishermen's unions, supported by the Indians.

In the 1930's, Alert Bay still possessed only a couple of miles of road, full of holes and ruts, but a visitor noted 16 cars and a motorbike. It was still an unsophisticated settlement, with washing flapping in the front yards of homes, and cows meandering down the road.

St. George's Hospital, which burned down in the 1920's and was replaced by a 27-bed building, was replaced again in 1947. A former Royal Canadian Air Force 60-bed hospital in Port Hardy was towed in 16 separate sections the 30 miles to Alert Bay and set up there. By the 1940's, Alert Bay had become an important fishing and logging centre, and reforestation was remedying some of the early forest destruction. Population had reached about 300 whites and 1000 Indians. In 1946 the settlement incorporated as a village. Three years later B.C. Packers donated property on which a community hall was built.

In the 1950's, British Columbia Airlines established a base at Alert Bay. A fire truck and the first island taxi began operating. In 1976 the bay's population was estimated at 760 and that of the whole island, including the reserve, at 1800. In 1978 the population was estimated at 2500, although residents say there has been an exodus from the island of both whites and Indians seeking higher-paying jobs. The canneries have gone, but there are boatyards and a net loft. Fishing is still the main industry. There are two old hotels, a bank, a movie theatre, and a small modern library and museum next to the Indian graveyard with its colourful grave totems. A triangular ferry system runs between Port McNeill, Alert Bay and Sointula.

The village still tends to cluster close to the water, but cross streets now extend up the hill for a few blocks, ending at the big field, in the centre of which the Indian community's longhouse stands. Tourists crowd in here when cruise ships dock at Alert Bay, watching the theatrical displays that are a gentle version of the Kwakiutl ceremonies of the past. Outside, they gaze upward at the

"world's tallest totem", 173 feet above ground, held upright by radiating guy wires. The immense red cedar was
donated by the Rayonier Company and was carved by
Benjamin Dick, his son Benjamin, Adam and William
Matilpi and Mrs. Billy Cook. One of the greatest attractions of the village is still the spectacular view of sea and
snow-capped mountains seen from its shores.

Indian and white children all attend a modern school
on the hillside, and the big Indian residential school is
now the Kwawkewlth Marine College, run by the reserve, open to all races and giving, among other subjects,
instruction in the carving of totem poles and other Nimpkish arts. Mungo Martin, the famous totem pole carver, is
from Alert Bay. Doug Cranmer, son of Christmas Tree
potlatcher Dan Cranmer and brother of the present Chief
Councillor, recently returned to Alert Bay after a 20-year
absence, during which time he worked at the University
of British Columbia, carving totem poles for its Indian
village. He has taught at the Kwawkewlth Marine
College.

According to Doug Cranmer, the Indians of Alert Bay
have no pressing racial problems today. Schools are integrated, a number of the Nimpkish have entered profes-
sions and, as in Cape Mudge, there has been a good deal
of intermarriage between whites and Indians over the
years.

Dan Cranmer's widow points out with pride that she
comes from Fort Rupert, where Kwakiutl tribes grouped
themselves around the old Hudson's Bay fort when it was
first built, and gave great potlatches to establish their relative rankings. The term Kwakiutl was orginally applied
only to the Fort Rupert tribes, who held top rank among
other tribes due to the spectacular potlatches they held
after the establishment of the fort in 1849 made
European goods available to them. Both Dan Cranmer's

son and his widow say the private Indian potlatches are still held to establish prestige and involve great expense, even though they no longer equal the huge gathering of Kwakiutl chiefs at the Christmas Tree Potlatch that resulted in the loss of the Nimpkish tribal treasures.

CHAPTER 12

Malcolm Island

"Oh Malcolm Island, Sointula,
Our home of peace and happiness,
Why ever were you created?
That is only known to him.
Of all that he has created upon earth
You have been granted to us.
To us you are precious
For you give us peace."
    -Matti Kurikka in *Aika*, May 16, 1902

Throughout the centuries, men have dreamed of Utopias; some have written books about their visions of an ideal community; some have tried to form one. In modern times, a dream most hopeful at the start and most disillusioned towards its close was the Finnish settlement of Sointula on Malcolm Island. Yet, however one ranks its success or failure, the story of Sointula's early days remains a moving tale of high ideals, arduous toil and heartbreaking misadventure.

It all began in the coal mines of North Wellington, near Nanaimo, around 1890. Finnish immigrants had settled in Nanaimo and found work in the mines, but conditions were bad and morale was low. Few safety precautions were taken; there had been an explosion in the mine in which many were killed; wages were poor, and there had been a labour strike. Then the mine owner, Premier James Dunsmuir, began mining operations at Extension, 12 miles from North Wellington. The Finns dismantled their houses and moved them to the new location, only to find shortly afterwards that Dunsmuir now planned to operate a mine at Oyster Bay and to establish a city there, renaming the spot Ladysmith. The miners were expected to move again, tearing down and rebuilding their houses once more. Lots were to be bought from Dunsmuir.

It was the last straw for the Finns. They were an idealistic group with ambitions for a better life. Sixteen of them had formed a Temperance Society called *Lännen Rusko* ("Western Glow.") They read widely, and most knew the works of Matti Kurikka, a passionate Finnish socialist, author, editor and playwright whose writings stressed power to the masses, the education and status of women, universal suffrage and the peace movement. At that time Kurikka was visiting Queensland, Australia, where he had hoped to rouse interest in establishing a Finnish community incorporating the Utopian vision as he saw it. His efforts there had failed, and when the Nanaimo Finns wrote to him, asking him to lead them to a promised land, he agreed to come to Canada.

Kurikka arrived in August of 1900, his passage fare of $125 paid by the miners. Full of enthusiasm, he prophesied: "In this colony, a high cultural life would be built, away from priests who have defiled the high morals of Christianity, away from churches that destroy peace, away from all the evils of the outside world."[1] Kurikka was made president of the group's Board of Directors, and he persuaded his friend A.B. Mäkëla to come from Finland to work with him as secretary. Cool and deliberate, Mäkëla was a contrast to the impulsive leader. Kurikka and Matti Halminen travelled to Victoria and came back with pamphlets and maps. They decided on Malcolm Island's 28,000 acres as a suitable site, sufficiently isolated from contaminating influences. Thickly forested with cedar, spruce and hemlock, the island for centuries had been the source of supply for Nimpkish Indian canoes, fishing tackle and household articles made from cedar and spruce.

Malcolm Island had received its name in 1846 from Commander George Gordon of H.M.S. *Cormorant,* honouring Admiral Sir Pulteney Malcolm. Napoleon said of the admiral, delegated to guard him on St. Helena: "Ah, there is a man with a countenance really pleasing, open, intelligent, frank and sincere.... His countenance bespeaks his heart, and I am sure he is a good man."[2]

While waiting for government approval, Kurikka started a paper, *Aika,* in Nanaimo, mainly to describe the future colony and urge others to join them. Finally, rules for the colony were drawn up by the Minister of Lands, and the agreement between the government and the Kalevan Kansa Colonization Company Ltd. was signed in the Finnish church.

The land was to be divided into 80-acre lots, one man to each 80 acres, ownership papers to be granted after seven years, provided the land was built on and improved up to $2.50 per acre. To qualify, the Finns were to become British subjects, each must possess a copy of the colony rules and sign an agreement to keep the regulations, the children were to attend an English school to be built by the government, while the settlers in turn were to build all roads, bridges and wharves. A yearly report was to be sent to the government. There would be no taxes imposed for the seven years except for poll tax of $3, and if the colony was successful, the government would grant them more land at the end of the seven-year period. Membership shares in the colony were sold for $200 a share, but those without money could exchange labour for shares. The colony was to be independent and self-sustaining through logging, fishing and agriculture.

Some years before, Malcolm Island had been the scene of another Utopian venture, when a preacher named Spencer had attempted to start a settlement of the Christian Utopian Society there. (The 1891 directory lists "Co-operative Colony, Malcolm Island".) Poverty and discord had brought this experiment to an early end. Charles William Cicero, rancher, Malcolm Island, is listed for 1898 only, in the Voters List. A Dane from San Francisco, E. Elliman, had also sought a peaceful life of seclusion on the island, where he raised chickens and vegetables. Indians found him dead a few months before the first party of Finns arrived.

The first Finnish settlers left Nanaimo on December 6, 1901, in a sailboat owned and captained by Johan Mikkelson. Theodore Tanner was the leader of the group, which included Kalle Hendrickson, Otto Ross and Malakias Kytomaa. A mishap occurred on the journey, when a shotgun was accidentally fired and Mikkelson was struck on the hand, later losing a thumb. Sailing and rowing, the men managed to reach Alert Bay, where Mikkelson was left to board the Nanaimo steamer when it docked there. The others reached Malcolm Island on December 15 and anchored at Rough Bay. They spent the first night on the boat, then moved into the shack that had belonged to the ill-fated Christian Utopians. That first year they lived in rough, little, two-room shacks among the ugly black stumps, the aftermath of logging.

The second group of settlers arrived in January of 1902, 14 men and a woman, led by Matti Halminen. Mrs.

Wilander, who had come from New York with her husband, was the first woman on the island and acted as cook for the men. The Wilanders lived in the old Spencer shack, with a hastily-constructed sauna attached. The trip up, on a sailboat bought to bring supplies to the island, had been frightening; none were experienced sailors and twice they were almost shipwrecked. Two men, Kilpelainen and Saarikoski, had been sent to the island to build a large log cabin to house families. (This later became the home of John Anderson and his family for many years.) The men worked steadily at felling trees, clearing brush and building log cabins, and the women carried coffee out to the workers in the afternoons for their brief period of rest. In the spring more people, some from the United States and some straight from Finland, arrived on the island. Dr. Oswald Beckman came from Astoria, Oregon, to join the settlement, so they were assured of medical care.

In June of 1902, it was decided to hold a mammoth celebration to advertise the settlement and plan the policy of the colony. The *Capilano* was chartered to bring up a large group from Nanaimo and surrounding areas, and the ship arrived at the dock with her decks crowded with men, women and children eager for their first glimpse of the Finnish Utopia.

Near the large log house, a roomy building called Cedar Hall had been erected to shelter the visitors. Close by, a big cedar stump served as a rostrum. With over a hundred members gathered to discuss the policies of the colony, it was decided to start logging operations at once behind Mitchell Bay, using the small steam donkey engine already on hand. The location for a future sawmill was decided upon, and work groups were organized, each with a leader who would be responsible for tools, boats or other equipment needed. The work day in those times was nine hours, but the colony decided on an eight-hour day after lengthy discussion. A townsite of half-acre lots was planned, and the name of the colony was officially changed from *Koti* to *Sointula*, meaning harmony.

Most of the visitors stayed behind when the *Capilano* left on her return trip. All went well during the pleasant summer months. The new settlers lived in tents while the weather was warm, but as winter set in the colonists realized they should have provided sufficient housing before encouraging new settlers to come. The log house and Cedar Hall could not begin to supply shelter for the many eager Finns who flocked to the island of harmony.

Now problems beset the colony. Debts amounted to $1300. The workers in the logging camp were inexperienced in the woods; many were tradesmen,—tailors, shoemakers and carpenters. The steam donkey was too small to handle huge logs properly; the price for logs was low and the cost of towing them to Vancouver was high. A salmon drag-seine was bought, and a steamboat, the *Vinetta*, but fish prices were also low, seven cents a fish for sockeye. The Alert Bay cannery had a monopoly for fish seven miles each way from the mouth of the Nimpkish. The Finns were given rights to build a cannery at Knight Inlet, but were unable to finance it due to their debts and other heavy expenditures.

Milk was needed for the children. The women picked salal berries in the summer, put them through a cider press and bottled the juice as a substitute, but finally cows were purchased in Nanaimo and shipped up on the *Capilano* in August of 1902. The cows arrived on the vessel, but so also did 20 more adult settlers and as many children, adding to the housing problem. Matti Kurikka was still writing with enthusiasm about Sointula and the need for settlers, in his paper *Aika*. In December of that year, five families with a total of 15 children arrived from Dakota, bringing with them four work horses.

There was no shelter or good pasture land for the animals, so they were taken to Vancouver Island where there were some old sheds standing on cleared land. Feed had to be taken across to them, and milk from the cows ferried back to Malcolm Island. In time, a large barn and loft were built on Sointula, and hay was brought from Wakeman Sound and other inlets by barge, towed by the *Vinetta*. Another large building of three storeys was constructed for the new families, with a big brick oven for heating and cooking. The top floor was set aside for a meeting hall.

Many settlers had been unable to pay cash for their membership fees and paid instead with labour. The financial situation of the colony was becoming critical, debts had climbed to $10,000 and Kurikka had great difficulty arranging further credit for the various improvements. At the end of January, 1903, when Kurikka returned to Sointula, a Mr. Bell came with him, sent by the creditors to investigate the logging and other work projects.

A meeting of members was called on the evening of January 29, and nearly all were present in the top floor of the big three-storey building. Children were brought along, and left in the lower rooms. The members were in the midst of a vehement discussion of Kurikka's report when someone screamed "Fire!"

Smoke and flames filled the narrow stairway and blocked escape. Some jumped from the windows, and those who were overcome by smoke were dropped down to them. Parents ran through the smoke-filled corridors of the lower floors, searching for their children. In all, two women, one man, and eight children perished in the fire. Some settlers who survived were badly burned, or injured from the leap from the third-storey windows. A.B. Mäkelä, the board's secretary and later the community's Justice of the Peace, was badly burned on face and hands and for a time it was feared his eyesight would be lost.

Word was sent to Alert Bay; police arrived, and an inquest was held.

The fire had begun near the brick oven, probably in the heating pipes. Cracks in the boards of ceiling and walls had allowed the fire to spread swiftly and fiercely. There was no sign of arson, but some discontented souls spread the rumour that Kurikka and Mäkëla started the fire to burn up their record books and remove evidence of embezzlement. Kurikka and Mäkëla disagreed about Kurikka's demand for the expulsion of two of the settlers and Kurikka won out after threatening to resign. Discord was creeping into the Finnish Utopia.

Nevertheless, the people continued to labour hopefully, erecting houses for the families and starting construction of a building for the newspaper *Aika*. It was decided to mortgage $20,000 worth of colony property to raise a $10,000 loan to pay off debts. There was an abundance of timber on the island and fish in the sea, but prices for lumber and salmon were still very low. Shoemakers and tailors found that shipping supplies from the towns for their trade resulted in higher costs to the buyer than were charged for factory-made articles. The brick factory and blacksmith shop were of great use to the community but made no profit for the colony. Only the sawmill produced a profit, and a new building, machinery and a larger towboat for the lumber were bought.

At the beginning of 1903 there had been 193 individuals on the island, including children; by December there were 238. Ten children had been born during the year. There had been talk for some time of building a community home for the children, where they could be left while the mothers worked. Motherhood, Kurikka believed, did not guarantee that mothers were suitable for child rearing. The experiment was tried out in 1904, but the mothers decided they preferred to raise the children themselves at home. A school with an English teacher,

Miss Cleveland, was opened that year, under an agreement on the part of the government to give the colony full rights to 640 acres in the townsite area providing the settlers built the $2,000 school.

Hope for a harmonious Utopia had not yet died. Despite their financial worries, the colonists remembered Kurikka's vision of a "high cultural life." They gave plays, held discussion groups, organized a band and choir. The emotional Kurikka wanted harmony in all things, including the music. Elizabeth Lempi Guthrie remembers how he would tear at his long, curly locks and cover his eyes when he heard a sour note, but says he was quick with praise when it was due.[3] A library was started, the first books coming from a group of Finnish people in Australia who had an organization for progressive ideas. But Kurikka's wish to be "away from all the evils of the outside world" could not be fully realized. The little socialist colony was dependent on the outside world of capitalism for survival.

That fall of 1904 saw the first of two crushing blows that preceded the liquidation of the colony. Kurikka was in Vancouver when bids were being sought for the building of bridges over the Capilano and North Seymour Rivers in North Vancouver. He placed a $3,000 offer for the contract in the name of the Kalevan Kansa Colonization Company Ltd., with a down payment of $150 on the bid.

Many in Sointula felt that the bid was unwise and that the $150 should be forfeited and the offer withdrawn, but Kurikka persuaded the colony that the work would lead to other more profitable contracts in the city, and that the abundance of timber on Malcolm Island would ensure success. His estimate proved to be woefully below the cost of the construction. He had failed to include cost of the foundation work, and far too little had been allowed for bolts, ties, and nails. Building material had to be hauled a mile up from shore to site, and a team of horses was bought for this. Men laboured without pay at the site and on the island for four months, and hundreds of thousands of feet of Malcolm Island's best timber was sacrificed. Many members of the colony left the island at this time in bitterness and despair.

Now the colony began to turn against its leader, both for his financial mistakes and for some of his ideas which were not endorsed by all the members. Kurikka had been advocating free love in his newspaper *Aika*. "Marriage and love are two different things, just as the church and truth are two different things," Kurikka said. His reason for the abolition of marriage was the servitude and lack of financial rights and freedom that married women often endured in those days, but readers in the outside world misinterpreted his motives. Many of the Sointula women, contentedly married, were distressed that Kurikka's beliefs were taken as their own. When Kurikka returned from his disastrous bridge contract, the colony was split between those who supported his ideas and those who were opposed. In the end, Kurikka resigned from the colony and left Sointula. Half the members went with him. Not all were his supporters. Many had lost everything they possessed in the bridge building and wanted only to escape from the island that meant to them no longer harmony, but only discord and failure. There were also those whom Mäkëla described as "windbags and fanatics, aggressive enough in spouting the principles of Utopian socialism but who preferred to leave the task of their realization to others."[4] Yet some wished to stay and work.

Those who were left were in a desperate position. They had lost half their labour force and were deeply in debt. Mäkëla suggested the colony should give a 50-year lease on 80-acre parcels of land to those who wished to remain in Sointula. This would assure them land, should the government decide to revoke the grant on the rest of the island. The last general meeting of the colony was held on February 5, 1905.

The second great blow occurred the following month. The sawmill was put to work sawing 2,000 feet of lumber, part of which went to the members, while the best 150,000 feet was to be marketed to pay interest on their mortgage and to buy desperately-needed shoes and clothing for the colony. Jarvinen, manager of the colony, was to deliver the lumber to the buyer, but he directed the tug to take it first to a location in North Vancouver near a lot where he planned to build. The skipper of the tug notified the colony's creditors and they promptly confiscated the lumber. The $3,000 the colony was counting on disappeared into the hands of the creditors.

There was no help for it now. The Kalevan Kansa Colonization Company was liquidated. Mäkëla arranged a loan to pay off the remainder of their debts, on condition that the land was returned to the government. Malcolm Island forests were sold for $5 per acre and all debts were repaid except for capital stock and the members' wages.

Penniless, their dream of owning a Utopian island shattered, the remaining settlers began the work of rebuilding their community. They worked in logging camps, fished and farmed, and slowly their hard work and faith in the future began to have results. They continued to meet in the Finnish Organization Hall, holding discussions, giving plays, and later presenting movies and holding dances.

In 1909, Andrew Anderson, in a letter to Premier Richard McBride, described the colony as consisting of 200 Finns, "just the people to make Malcolm Island one of the best settlements on the British Columbia Coast", living in "beautiful, neat homes."[5] They were still farming and owned cattle, though they had not yet developed hayfields on the island. According to some of the sea captains, the farmers were shipping out ten to 50 cases of eggs a week to both northern and southern communities. There were no horses on the island as there were no roads for travel, and this was one of their pressing problems because for half the year the weather was too rough to travel along their coastline in their little rowboats.

John Anderson, who had been a sailor in Finland, had rowed over to Sointula from Alert Bay in 1906 and arrived to find the settlers celebrating Midsummer Eve in the Finnish tradition around huge bonfires on the beach. The homesick young man was happy to hear Finnish spoken, and decided to join the colony even though at that time it meant a four-mile trip by rowboat to Alert Bay whenever groceries were needed. In 1909, the community started its co-operative store, first in a private house, and when this burned down with a loss of $20,000, the Sointula Co-operative Store replaced it in 1933. Everyone in the community was a part of the operation.

John Anderson noted that one great drawback to the choice of Malcolm Island was the lack of a good harbour; there is no sheltered cove along the coastline of the 15-mile-long island. With no roads in the early days and the sea the only highway, transportation ceased when summer westerlies or winter southeasters blew hard. Today, there is access to most of the island by public or logging roads, and many residents own cars.

The school was replaced in 1928 by a new, three-room building serving higher grades, and many children went on to enter professional careers. Virginia Johnston taught in the Sointula school in the 1920's and found many of the precepts of Kurikka still in force. She estimated that 98% professed no formal religion but there were regular meetings on Sundays to study nature or practise physical drills. Sointula was an orderly community. There was no policeman on the island, nor was there any need for one.

Commercial fishing soon became the main source of revenue, with the new generation turning from the soil to the sea for a livelihood. By 1948 there were 70 gillnetters, 12 deepsea trollers and several big seiners in the Sointula fleet. Boats increased in convenience and comfort as modern methods were invented. It was a Finn, Lauri Jarvis, who invented the fishboat drum on which nets could be hauled in with machinery instead of by backbreaking manual labour. Always a supporter of labour movements, Sointula became a strong local chapter of the United Fishermen's and Allied Workers' Union. Boatbuilding was carried on in Eino Tammi's shipyard for over 30 years, until in 1977 he sold out to Triangle Shipyard. John Anderson also managed a boatyard in Sointula.

By 1950 the British Columbia Power Commission brought in electricity from the Alert Bay plant, and telephones were installed by the North-West Telephone Company in 1956. Gone were the days of wood-burning stoves and the need to saw and chop beach logs for fragrant wood fires. With the influx of English-speaking residents, Sointula has ceased to be a strictly Finnish community. The co-operative store is still operating, and the big Finnish Organization Hall still looms up on the hillside by the fire station, but plays and discussions are seldom held, superseded by television. Kurikka founded the colony to be "away from churches that destroy peace", but recently a church was built "for those who wish to participate."

There are other traces to be seen of the old colony, in the little library that contains some of the original Finnish books, in abandoned farms and saunas, and in the neatly-kept graveyard where the hammer and sickle emblem appears on the earliest tombstones along with tributes to the industrious, inspired pioneers of Sointula's Utopia.

Epilogue

Through the journals of explorers and surveyors, through the tales of loggers and miners and the recollections of pioneer settlers, the history of the islands is unfolded. The aggressive Kwakiutl no longer menace voyagers in Discovery Passage, and their dances and potlatches, once outlawed by the government, are now advertised as attractions in government pamphlets for tourists. The fearsome reefs and rapids have been charted and the vagaries of the tidal currents have been recorded. The influx of settlers to some of the islands has slowed or ceased as the once-flourishing industries have dwindled. Many residents have succumbed to the lure of higher-paying jobs or easier living conditions in the towns and cities.

Like the boats anchored above the rapids, waiting for a

change of tide, the northern islands of the Inside Passage wait for the changes that lie ahead. There is still some logging and fishing, but gone for the present are those busy years of camps and canneries on so many of the islands. Gone also are the Union Steamships that played so large a part in early settlement and provided a link with the rest of the province.

The conservation of fish and forest practised today may bring back some of the activity in the course of time, though it may never reach the prolific state of a century ago when surveyor Richard Mayne said: "The seas and large inlets, the bays and rivers are literally alive with fish," and lumberjack Jack Hannahar said: "It will be many a day yet before the islands will be cleared of their timber."

Already, many of the islands depend largely on the tourist trade. Ahead for others may lie the same influx of tourists, as the southern Gulf Islands grow more and more crowded and the search continues for new scenes for holidays and homes. If not, these northern islands will remain a needed haven for those who seek escape from the noise and the increasing tensions of city life.

NOTES

Introduction: The Islands

1. Smith, Marcus, "Progress Report on the Surveys for 1872 in British Columbia" (Appendix E of the *Report of Progress on the Explorations and Surveys for the Canadian Pacific Railway, up to 1874* by Sandford Fleming, Engineer-in-Chief), Section: "Journey to Bute Inlet and Commencement of the Survey," p. 108.

Quadra Island

1. Millay, Edna St. Vincent, *There Are No Islands Any More*, June 1940

2. Vancouver, George, *A Voyage of Discovery*. v.1, entry for September 5, 1792.

3. Mayne, Richard C., *Four Years in British Columbia and Vancouver Island*. p. 176.

4. Fleming, Sandford, *Report of Progress on the Explorations and Surveys for the Canadian Pacific Railway, up to 1874*. p. 23.

5. Meany, Edmond S., *A New Vancouver Journal on the Discovery of Puget Sound by a Member of the Chatham's Crew*. Introduction, pp. 2 and 25.

6. Vancouver, George, *A Voyage of Discovery*. v.1, entry for September 5, 1792.

7. Newcombe, C.F. ed., *Menzies' Journal of Vancouver's Voyage, April to October 1792*. entry for July 13, 1792.

8. Curtis, E.S., *The North American Indian*. v.10, pp. 105, 106.

9. Mayne, Richard C., *Four Years in British Columbia*. p. 177.

10. Waite, Donald E., *The Langley Story*. pp. 17, 18.

11. Curtis, E.S., *The North American Indian*. v.10, Appendix: "Kwakiutl tribes."

12. *British Columbia Directory*. 1892, entry under "Valdes Island."

13. Pidcock, R.H., "Diary", entry for July 3, 4, 5, 1888.

14. Columbia Coast Mission, *The Log*, July 3, 1906, report by Mrs. B.E. Ward: "A Gala Day at Heriot Bay, Valdez Island."

15. Canada, Sessional Papers, 1885, *Report of the Department of Indian Affairs*, letter of Agent George Blenkinsop, dated June 4, 1884. Referring to the Laich-kwil-tach, who, he says, "are in every way superior to the other branches of the Kwawkewlth family," he continues, "They are now anxious to have a school opened at Cape Mudge village . . . unanimous in their desire for the education and general improvement of the young."

16. Pidcock, R.H., "Diary", entry for April 20, 1888.

17. Canada, Sessional Papers, 1895, *Report of the Department of Indian Affairs*, letter of Agent Pidcock, dated August 25, 1894.

18. Walker, A.K., "Indian Missionary Activity", n.p.

19. Canada, Sessional Papers, 1911, *Report of the Department of Indian Affairs*, letter of Agent William Halliday, dated March 31, 1910.

20. Letter in *Toronto Empire*, February 1893, headed "The Evil Potlatch."

21. *Colonist*, August 9, 1892.

22. Newspaper article: "Chief Billy Assu Tells of Early Days", in the Campbell River and District Museum, n.d. or source.

23. Manson, Michael, "Sketches from the Life of Michael Manson".

24. Interview with the author, 1976.

25. Interview with the author, 1975.

26. Curtis, E.S., *The North American Indian*, v.10, pp. 162, 163.

27. Interview with the author, 1976.

28. *Colonist*, March 2, 1905, letter to the editor, dated February 17, 1905 and signed "Just Cause, Cape Mudge."

29. *Colonist*, September 21, 1969, article by Helen A. Mitchell: "Pioneers of Granite Bay".

30. Antle, John, "Memoirs of John Antle", 1905–1936. *Lavrock* and *Laverock* are two spellings used interchangeably by Antle and other authors.
31. *Ibid.*
32. Columbia Coast Mission, *The Log.* July 3, 1906, report by Mrs. B.E. Ward: "A Gala Day at Heriot Bay, Valdez Island".
33. *Campbell River Courier.* November 12, 1976, article by C. G. Seymour Bagot: "Quadra Island: One Man's Memory".
34. Diary of George Rose, in possession of Mrs. George Rose of Francisco Point, Quadra Island.
35. Gould, Ed., *Logging: British Columbia's Logging History*; also *Colonist.* April 2, 1961, article by Maude Emery' "The Loggers Found Rich Gold Deposits".
36. Columbia Coast Mission, *The Log.* September-October 1944.
37. Columbia Coast Mission, *The Log.* 1907.
38. *Colonist.* November 20, 1903.
39. Antle, John, "Memoirs of John Antle".
40. Interview with the author, 1976.
41. 1951 Fieldnotes of Helen Codere in her *Daniel Cranmer's Potlatch:* (See McFeat, Tom, *Indians of the North Pacific Coast.*)
42. *Ibid.*
43. Halliday, William, *Potlatch and Totem.* p. 194.
44. Interview with the author, 1976.
45. Lyons, Cicely, *Salmon. Our Heritage.* p. 399.
46. Newspaper article: "Chief Billy Assu Tells of Early Days", in the Campbell River and District Museum, n.d. or source.
47. *Vancouver Sun.* January 19, 1972, p. 32.
48. *Province.* March 18, 1960, p.4.
49. Regional District of Comox-Strathcona, Planning Department, *Quadra Island Planning Study 1971.*
50. *Colonist.* June 20, 1974.

51. *Campbell River Up Islander*, July 21, 1976.
52. *Discovery Passage*, October 11, 1974.
53. *Ibid.*, July 6, 1972.

CORTES ISLAND

1. *Campbell River Courier*, 23 January, 1957, "Uncle John Manson Laid to Rest on Cortes", Provincial Archives, Victoria.
2. Barnett, Homer G., *The Coast Salish of British Columbia*, pp. 26, 27.
3. Newcombe, C.F., ed., *Menzies' Journal of Vancouver's Voyage, April to October 1792*, entry for July 2, 1792.
4. Vancouver, George, *A Voyage of Discovery*, v.1, entry for July 13, 1792.
5. *Colonist*, November 3, 1968, p. 11, article by Jack McQuarrie: "Few Indians Can Build Dugout Canoe".
6. *British Colonist*, September 15, 1869, letter to the editor, signed Abel Douglass, dated September 2, 1869, headed "The Dawson Whaling Party—The Perils of Whaling".
7. Manson, Michael, "Sketches from the Life of Michael Manson".
8. British Columbia, Sessional Papers, 1903, *Report on Agriculture*, letter of Mr. Nicholas Thompson of Whaletown, Cortez Island.
9. Columbia Coast Mission, *The Log*, 1908.
10. *Ibid.*, May 1934.

READ ISLAND

1. Columbia Coast Mission, *The Log*, January-February 1943, p. 184.
2. British Columbia, Public Inquiries Act, Report of the Commissioner, *The Forest Resources of British Columbia*, v.1.
3. Canada, Department of Mines, *Geology of the Coast and Islands Between the Strait of Georgia and Queen Charlotte Sound, B.C.*, by J. Austen Bancroft, Memoir No. 23, section under "Clays, Lime and Cement."

4. Letter from C. Redford, Fawn, B.C. to Robert Tipton, Surge Narrows, B.C., dated July 19, 1925, in the Campbell River and District Museum.

5. *Colonist*. September 25, 1977, article by Doris Davies: "Read Island School Teacher—1926 Style".

THE REDONDA ISLANDS

1. Vancouver, George, *A Voyage of Discovery*. v.1, entry for June 26, 1792, p. 322.

2. Newcombe, C.F. ed., *Menzies' Journal of Vancouver's Voyage. April to October 1792*. entry for June 27, 1792.

3. Wagner, Henry R., *Spanish Explorations in the Strait of Juan de Fuca*. extract from the *Diary of Galiano and Valdes*. p. 217.

4. Vancouver, George, *A Voyage of Discovery*. v.1, entry for July 13, 1792, p. 337.

5. *Ibid.*, p. 339.

6. Walbran, Captain John T., *British Columbia Coast Names*. entry under "Homfray Channel."

7. Canada, Department of Mines, *Geology of the Coast and Islands Between the Strait of Georgia and Queen Charlotte Sound. B.C.*, by J. Austen Bancroft, Memoir No. 23, section under "Building and Ornamental Stones."

8. Manson, Thomas, "Historical Geography of Redonda Bay".

9. *Ibid.*

STUART ISLAND

1. *A Narrative or Journal of a Voyage of Discovery to the North Pacific Ocean and Round the World. performed in the years 1791, 1792, 1793, 1794 and 1795 by Capt. George Vancouver and Lieutenant Broughton*. London, J. Lee, 1802, p. 48.

2. Vancouver, George, *A Voyage of Discovery*. v.1, entry for July 2, 1792.

3. Wagner, Henry R., *Spanish Explorations in the Strait of Juan de Fuca*. p. 275.

4. Espinosa y Tello, Jose, *Spanish Voyage to Vancouver Island and the Northwest Coast of America*, trans. by Cecil Jane, p. 65.

5. Wagner, Henry R., *Spanish Explorations*, p. 219.

6. Fleming, Sandford, *Report of Progress on the Explorations and Surveys for the Canadian Pacific Railway, up to 1874*, p. 22.

7. Smith, Marcus, "Progress Report on the Surveys for 1872 in British Columbia," (Appendix E of the *Report of Progress*, see above), July 11, 1872, p. 114.

8. *Ibid.*

9. Willcock, M.V., *Vancouver Sun*, 25 July, 1953.

MAURELLE ISLAND

1. Wagner, Henry R., *Spanish Explorations in the Strait of Juan de Fuca*, p. 275.

2. Canada, Sessional Papers, 1895, *Report on Agriculture*, letter of Tom Bell, headed "Valdez Island," June 3, 1895, to J.R. Anderson.

SONORA ISLAND

1. Newcombe, C.F. ed., *Menzies' Journal of Vancouver's Voyage, April to October 1792*, entry for July 2, 1792.

2. Smith, Marcus, "Progress Report on the Surveys for 1872 in British Columbia" (Appendix E of the *Report of Progress*), section on "Waddington Harbour to Vancouver Island," p. 153.

3. Columbia Coast Mission, *The Log*, May-June 1946; Alan Greene writes in more detail of Hiram Corn.

THE THURLOW ISLANDS

1. Vancouver, George, *A Voyage of Discovery*, v.1, entry for July 16, 1792.

2. Rushton, Gerald, *Whistle Up the Inlet*, p. 40.

3. British Columbia, Sessional Papers, 1898, *Report of the*

Minister of Mines. Nanaimo District: Phillips Arm and Shoal Bay.

4. Quoted in Pethick, Derek, *S.S. Beaver. The Ship That Saved the West.* p. 104.

CRACROFT, MINSTREL AND HARBLEDOWN ISLANDS

1. Columbia Coast Mission, *The Log,* July-September 1960, p. 9.

2. Curtis, E.S., *The North American Indian.* v. 10, p. 294.

CORMORANT ISLAND

1. British Columbia, Sessional Papers, 1894–95, *Crown Land Surveys.* Survey of Toba Inlet, Powell Lake and Kingcome Inlet, Report of A.F. Cotton, P.L.S., January 7, 1895.

2. Pidcock, R.H., "Diary", 1888.

3. Boas, Dr. Franz, *The Indians of British Columbia.* American Geological Society, v. 28, pp. 229–243.

4. Pidcock, R.H., "Diary", 1888.

5. Canada, Sessional Papers, 1889, *Report of the Department of Indian Affairs.* letter of R.H. Pidcock, September 3, 1888.

6. *Discovery Passage.* May 1977, pp. 2, 3, article by Joy Inglis.

MALCOLM ISLAND

1. Anderson, Aili, *History of Sointula.* p. 2.

2. Walbran, Captain John, *British Columbia Coast Names.* entry under "Malcolm Island."

3. *Ladysmith Chronicle.* January 11, 1962, article by Mrs. E.L. Guthrie.

4. Kolemainen, John I., "Harmony Island—A Finnish Utopian Venture in British Columbia," *British Columbia Historical Quarterly.* v.5, n.2, April 1941, p. 120.

5. Provincial Archives of British Columbia, Premier: Correspondence Inward, no. 280/09 (official), Andrew Anderson to Richard McBride, May 25, 1909.

EPILOGUE

1. From the essay *The Wave of the Future.* 1940, by Anne Morrow Lindbergh.

SOURCES

BOOKS

Akrigg, G.P.V. and H.B. Akrigg. *1001 British Columbia Place Names.* Vancouver, Discovery Press, 1969.

Anderson, Bern. *Surveyor of the Sea.* Seattle, University of Washington Press, 1960.

Bancroft, Hubert H. *History of the Northwest Coast.* vol. 1. San Francisco, Bancroft, 1884.

Barnett, Homer G. *The Coast Salish of British Columbia.* Eugene, Oregon, University of Oregon Press, 1955.

Boas, Franz. *Contributions to the Ethnology of the Kwakiutl.* New York, Columbia University Press, 1925.

_____. *The Social Organization and the Secret Societies of the Kwakiutl Indians.* Washington, Government Printing Press, 1897.

Codere, Helen. *Fighting With Property.* New York, Augustin Press, 1950.

Curtis, Edward S. *The North American Indian.* vols. 9, 10. New York c. 1911, Johnson Reprint 1970.

Dawson, George. *Report on a Geological Examination of the Northern Coast of Vancouver Island and Adjacent Coasts.* Montreal, Dawson Bros., 1887.

Dawson, Will. *Coastal Cruising.* 3rd rev. ed. Vancouver, Mitchell Press, 1973.

Driver, Harold E. *Indians of North America.* Chicago, University of Chicago Press, 1969.

Drucker, Philip. *Indians of the Northwest Coast.* New York, McGraw-Hill, 1955.

_____. *The Native Brotherhoods; Modern Intertribal Organization on the Northwest Coast.* Washington, Smithsonian Institution, Bulletin 162, 1958.

_____ and R.F. Heizer. *To Make My Name Good; a re-examination of the Southern Kwakiutl potlatch.* Berkeley, University of California Press, 1967.

Espinosa y Tello, José. *Account of the Voyage Made by the Schooners Sutil and Mexicana in the Year 1792.* Trans. by G.F. Barwick, London, 1911. From 1802 ed., Madrid, Royal Printing Office.

_____. *Spanish Voyage to Vancouver Island and the Northwest Coast of America.* Trans. by Cecil Jane. London, Argonaut Press, 1930. From 1802 ed., Madrid, Royal Printing Office.

Forester, Joseph E. *Fishing: British Columbia's Fishing History.* Saanichton, B.C., Hancock House, 1975.

Fry, Alan. *How a People Die.* Garden City, N.Y., Doubleday, 1970.

Gosnell, R.E. *Yearbook of British Columbia and Manual of Public Information.* Victoria, 1897.

Gould, Ed. *Logging: British Columbia's Logging History.* Saanichton, B.C., Hancock House, 1975.

Greene, Ruth. *Personality Ships of British Columbia.* Vancouver, Marine Tapestry, 1969.

Gunther, Erna. *Indian Life on the Northwest Coast of North America.* Chicago, University of Chicago Press, 1972.

Halliday, William. *Potlatch and Totem.* London, Dent, 1935.

Hodge, F. *Handbook of Indians of Canada.* Ottawa, Parmelee, 1913.

Howay, F.W. and E.O.S. Scholefield. *British Columbia from Earliest Times to the Present.* Vancouver, Clarke, 1914.

La Violette, F.E. *The Struggle for Survival.* Toronto, University of Toronto Press, 1961.

Lyons, Cicely. *Salmon, Our Heritage.* Vancouver, Mitchell Press, 1969.

McCurdy, H.W. *The H.W. McCurdy Marine History of the*

Pacific Northwest. Ed. by Gordon Newell. Seattle, Superior Publishing, 1977.

McFeat, Tom. *Indians of the North Pacific Coast.* Toronto, McClelland and Stewart, 1966.

Marshall, James S. and Carrie Marshall. *Vancouver's Voyage.* Vancouver, Mitchell Press, 1969.

Mayne, Richard C. *Four Years in British Columbia and Vancouver Island.* London, John Murray, 1862.

Mitchell, Helen A. *Diamond in the Rough.* 1966.

Morton, James W. *The Enterprising Mr. Moody, the Bumptious Captain Stamp.* North Vancouver, Douglas and McIntyre, 1977.

Newcombe, C.F. *The First Circumnavigation of Vancouver Island.* Victoria, Cullin, 1914.

———· (ed.) *Menzies' Journal of Vancouver's Voyage, April to October, 1792.* Victoria, 1923.

North, George. *A Ripple, a Wave: The story of union organization in the B.C. fishing industry.* Rev. and ed. by Harold Griffin. Vancouver, Fisherman, 1974.

Pethick, Derek. *S.S. Beaver: The Ship That Saved the West.* Vancouver, Mitchell Press, 1970.

Phillips, Paul. *No Power Greater; a century of labour in B.C.* Vancouver, Federation of Labour, Boag Foundation, 1967.

Richards, Sir George Henry. *The Vancouver Island Pilot.* Great Britain, Hydrographic Office, 1864.

Rushton, G.A. *Whistle Up the Inlet.* Vancouver, J.J. Douglas, 1974.

Smith, Marcus. "Report on the Surveys for 1872 in British Columbia" (Appendix E of the CPR *Report of Progress on the Explorations and Surveys up to January 1874.*) Ottawa, Maclean, Roger & Co., 1874.

Spradley, James P. *Guests Never Leave Hungry.* New Haven, Yale University Press, 1969.

Thornton, Mildred V. *Indian Lives and Legends.* Vancouver, Mitchell Press, 1966.

Vancouver, George. *A Voyage of Discovery,* vol. 1. London, G.G. & J. Robinson, 1798.

Wagner, Henry R. *Cartography of the Northwest Coast of America to the Year 1800.* Berkeley, Ca., University of California Press, 1937.

_____. *Spanish Explorations in the Strait of Juan de Fuca.* Santa Ana, Ca, Fine Arts Press, 1933.

Waite, Donald E. *The Langley Story.* Don Waite Publishing, 1977.

Walbran, Captain John T. *British Columbia Coast Names.* Vancouver, J.J. Douglas, 1971.

PAMPHLETS

Anderson, Aili. *History of Sointula.* Supplement by Mrs. Aini Tynjala. 1969.

British Columbia Teachers Federation. *To Potlatch or Not to Potlatch.* Lesson Aid no. 2011, 1972.

Canada. Department of Mines. *Geology of the Coast and Islands Between the Strait of Georgia and Queen Charlotte Sound, B.C.,* by J. Austen Bancroft. Memoir No. 23. Ottawa, Government Printing Bureau, 1913.

Cherukapalle, Nirmala Devi. *Indian Reserves as Municipalities, problems and policies.* Papers on Local Government, vol. 1, no. 3. Vancouver, Centre for Continuing Education, University of British Columbia, 1972.

Duff, Wilson. *The Indian History of British Columbia.* vol. 1: *The Impact of the White Man.* British Columbia Provincial Museum, 1964.

Duncan, Frances. *The Sayward-Kelsey Bay Saga.* Courtenay, Argus, 1958.

Healey, Elizabeth. *History of Alert Bay and District.* Alert Bay Centennial Committee, n.d.

Nelson, Denys. *Fort Langley 1827-1927.* Vancouver, Art, Historical and Scientific Association of Vancouver, 1927.

Meany, Edmond S. *A New Vancouver Journal on the Discovery of Puget Sound by a Member of the Chatham's Crew.* Seattle, Washington, 1915.

Regional District of Comox-Strathcona. Planning Department. *Quadra Island Planning Study 1971.*

PUBLISHED ARTICLES

Boas, Dr. Franz. "The Indians of British Columbia". *Bulletin of the American Geological Society*. vol. 28, pp. 229-243.

Kolemainen, John I. "Harmony Island: A Finnish Utopian Venture in British Columbia." *British Columbia Historical Quarterly*. vol 5, no. 2, April 1941.

Longstaff, F.V. "Captain George Vancouver 1792-1942, a Study in Commemorative Place Names". *British Columbia Historical Quarterly*. vol. 6, 1942.

Meade, E.F. "A Euclataw Chief". *The Beaver*. Winter 1965.

_____. "The Mute Ghosts of Cape Mudge". *The Beaver*. Autumn 1962.

Taylor, Herbert C. and Wilson Duff. "A Post-contact Southward Movement of the Kwakiutl". *Research Studies*, State College of Washington, vol. 24, 1956.

Wagner, Henry R. and C.F. Newcombe, eds. "Extract from the Journal of Jacinto Caamano", trans. by Harold Grenfell, R.N. *British Columbia Historical Quarterly*. July 1938.

Wilson, Donald. "Matti Kurikka: Finnish Canadian Intellectual". *British Columbia Studies*. no. 20, Winter 1973-74.

UNPUBLISHED MATERIAL

Antle, John. "Memoirs of Dr. John Antle 1905-1936." Typescript. Northwest Room, Vancouver Public Library.

Manson, Michael. "Sketches from the Life of Michael Manson". Typescript. British Columbia Provincial Archives, Victoria.

Manson, Thomas. "Historical Geography of Redonda Bay". Typescript. 1975. British Columbia Provincial Archives, Victoria.

Pidcock, George. "Diaries". 1902, 1903, 1908-11, 1913-23. Holograph. British Columbia Provincial Archives, Victoria.

Walker, Agnes Knight. "Indian Missionary Activity on Quadra Island". Holograph. British Columbia Provincial Archives, Victoria.

MAGAZINES

The Beaver. Autumn 1962, Winter 1965.

Columbia Coast Mission. *The Log.* Vancouver, B.C., 1906–1909, 1930–1973.

Pacific Yachting. 1976, 1977.

NEWPAPERS

British Colonist. Victoria

Campbell River Courier

Campbell River Up Islander

Comox Free Press

Discovery Passage

The Native Voice

UF & AWU Fisherman

The Province. Vancouver

Vancouver Sun

The Colonist. Victoria

Victoria Daily Times

GENERAL REFERENCE

British Columbia Directory. 1892–1940.

British Columbia. Public Inquiries Act. Report of the Commissioner. *The Forest Resources of British Columbia.* vol. 1, 1956.

British Columbia. Sessional Papers. Annual Reports.

Canada. Sessional Papers. Reports of the Department of Indian Affairs.

INDEX